NAPOLEON'S CAVALRY: A KEY ELEMENT
TO DECISIVE VICTORY

A thesis presented to the Faculty of the U.S. Army
Command and General Staff College in partial
fulfillment of the requirements for the
degree

MASTER OF MILITARY ART AND SCIENCE
Military History

by

THOMAS A. SHOFFNER, MAJ, USA
B.S., United States Military Academy, West Point, New York, 1990

Fort Leavenworth, Kansas
2002

Approved for public release; distribution is unlimited.

ABSTRACT

NAPOLEON'S CAVALRY: A KEY ELEMENT TO DECISIVE VICTORY,
by MAJ Thomas A. Shoffner, 107 pages.

Napoleon's rise to power in the late eighteenth century occurred at a time when the structure of most European armies was based on the paradigm army of Frederick the Great. Napoleon, however, changed all of this and in a few short years transformed the French army into the most powerful force on the continent of Europe. During the period of 1805 to 1813, Napoleon's army had no equal with regard to operational effectiveness. Speed and positioning of forces were the two main characteristics that made the French army so successful. These same two characteristics were also inherent to French cavalry units. Thus, the central research question is: What influence did cavalry have upon Napoleon's operations? To facilitate this study, two campaigns were examined that illustrate cavalry's impact on Napoleon's operations. The first campaign was the Jena Campaign of 1806; the second was the Saxony Campaign of 1813. The Jena Campaign demonstrated that with the employment of sufficient and well-trained cavalry, Napoleon could render his victories decisive through the complete destruction of the enemy army. Conversely, the Saxony Campaign demonstrated that without the effective employment of sufficient and well-trained cavalry, Napoleon could not obtain the complete destruction of the enemy army and thus, his victories were hollow, or at best Pyrrhic. Therefore, based on the analysis of these two campaigns, this study has concluded that Napoleon's cavalry was a key element for Napoleon achieving complete destruction of the enemy army, thus rendering his victories decisive.

LIST OF ILLUSTRATIONS

Figure	Page
1. Jena Campaign. Situation as of 6 October 1806	27
2. Jena Campaign. Situation as of Noon 10 October 1806	30
3. Jena Campaign. Situation as of 6 P.M., 12 October 1806	33
4. Battles of Jena and Auerstadt. Situation as of Midnight, 13 October 1806	36
5. Battle of Jena. Situation as of 10 A.M., 14 October 1806	39
6. Battle of Jena. Situation as of 2 P.M., 14 October 1806	42
7. Battle of Auerstadt. Situation as of 10 A.M., 14 October 1806	44
8. Battle of Auerstadt. Situation as of 2 P.M., 14 October 1806	46
9. Jena Campaign. Pursuit to the Oder. 15 October to 1 November 1806	49
10. Battle of Lutzen, 2 May 1813	70
11. Battle of Bautzen, 20-21 May 1813	74
12. Battle of Leipzig, 19 October 1813	85

LIST OF TABLES

Table	Page
1. Cavalry Missions and Capabilities of the Early Nineteenth Century	17
2. French Order of Battle for the Jena Campaign, October 1806	24
3. French Order of Battle for the Saxony Campaign, October 1813	65

ABBREVIATIONS

BFV	Bradley fighting vehicle
CSS	combat service support
FM	field manual
FCS	future combat system
HIC	high intensity conflict
HMMWV	high-mobility multipurpose wheeled vehicle
IBCT	interim brigade combat team
LAV	light armored vehicle
LIC	low intensity conflict
LOC	lines of communications
MOOTW	military operations other than war
MRC	major regional conflict
NBC	nuclear, biological, and chemical
RSTA	reconnaissance, surveillance and target acquisition
TTP	techniques, tactics, and procedures
UAV	unmanned aerial vehicle
US	United States

CHAPTER 1

INTRODUCTION

Napoleon's rise to power in the late eighteenth century occurred at a time when the structure of most European armies was based on the paradigm army of Frederick the Great. Napoleon, however, changed all of this and in a few short years transformed the French army into the most powerful force on the continent of Europe. During the period of 1805 to 1813, the French army had no equal with regard to operational effectiveness. The two main characteristics of his army were speed and positioning of forces--two traits that are inherent to cavalry units. These two traits have recently gained renewed interest of the United States Army as it goes through a process of transformation to maintain strategic relevance during a period of redefinition of national strategy.

Since the end of the Vietnam War, the forces of the US Army have predominantly focused on the extreme ends of the spectrum for military operations--light infantry units for low intensity conflicts (LICs) and military operations other than war (MOOTW), and heavy mechanized and armored forces for high intensity conflict (HIC), and major regional conflicts (MRCs). For today's emerging strategy, the heavy forces have been determined to be too heavy for rapid deployability; and the light forces, while rapidly deployable, are lacking in lethality and survivability. To mitigate this legacy force capability gap the Army is currently undergoing a transformation process in order to develop rapidly deployable forces that span the full spectrum of military operations. This transformation process primarily deals with exploring new doctrine and equipment. Of recent interest is the introduction of the Interim Brigade Combat Team (IBCT) that is scheduled to be fielded with a wheeled, Lightly Armored Vehicle (LAV).

Concomitantly, the Army is undergoing a review and examination of doctrine and techniques, tactics, and procedures (TTPs) needed to properly employ this new equipment to its maximum effectiveness. The IBCT is to have the speed and ability to maneuver that far exceeds that of the dismounted infantryman and be much more deployable than the heavy forces. While the family of future combat systems (FCS) is being developed, this interim force is to have strategic relevance and provide important insights into the doctrine and TTP needed for the objective force. This transformation process provides the potential for the Army to align its equipment and doctrine with one that might closely relate to Napoleon's three classes of cavalry and to the methods he used to employ them.

Concurrent with the fielding of the IBCT, the US Army has recently published its updated operations manual: FM 3-0, *Operations*. While FM 3-0 continues to address the three levels of war (strategic, operational, and tactical), it has also categorized Army operations into three types: decisive, shaping, and sustaining. These two items, the levels of war and the types of operations, provide a relevant construct from which to compare and contrast the effectiveness of Napoleon's cavalry.

In light of the Army's transformation and the release of its capstone operations doctrine, FM 3-0, this study will examine the influence and impact of cavalry upon Napoleon's operations in an effort to gain further insights and lessons for not only the IBCT, but also for future mounted operations. To facilitate this study, two campaigns were selected that illustrate the cavalry's impact on Napoleon's war effort. The first campaign to be examined is the Jena Campaign of 1806; the second is the Saxony Campaign of 1813. By studying these two campaigns, a series of questions will be answered. How did Napoleon organize and employ his cavalry with regard to the three categories of operations outlined in FM 3-0 (decisive, shaping, and sustaining operations)? Did Napoleon employ his cavalry any differently depending on the particular level of war he was focused on? What are the insights from these two campaigns that may be applied to mounted operations today? And finally, are there any operational or tactical techniques that may be applied to today's battlefield, or that should be considered for design of today's force structure?

The study will begin with an examination of the three classes of cavalry (light, medium, and heavy cavalry) and how each was organized, equipped, and employed. The analysis will attempt to discover if any particular considerations were given to certain types of unit designs and their battlefield impact. This chapter will also consider the level(s) of war the different cavalry units made their most significant contributions. Of particular interest will be any battlefield insights that can be applied to today's mounted forces.

My study will then examine the impact cavalry, or lack of cavalry, had on the outcome of the Jena and Saxony Campaigns. There are two reasons for selecting these campaigns. First, each occurred during the height of Napoleon's military success. Second, both campaigns illustrate either the successful employment of cavalry (before, during, and after an operation), or the impact the lack of cavalry had on the outcome of a campaign. The two campaigns also allow for the examination of the employment of cavalry at each of the three levels of war as well as the three types of military operations as referenced in FM 3-0. Realizing there is insufficient time to thoroughly examine all of

Napoleon's campaigns within the constraints of this paper, this study has intentionally exclusively on the selected campaigns of Jena and Saxony.

The two campaigns do provide a sufficient example to compare and contrast the employment of cavalry in two distinct situations. The Jena campaign provides successful examples of Napoleon's speed and positioning of forces, as well as a classic example of pursuit that led to the defeat of the Prussian Army. The Saxony campaign provides examples of the impact of insufficient cavalry forces on a campaign.

The methodology selected to analyze these campaigns will address how each type of cavalry was employed at each level of war for the operation; and the cavalry's contribution to the decisive, shaping, and sustaining operations that led to the outcome of the campaign.

The desired end state for this study will be to obtain and comprehend the insights demonstrated during the Jena and Saxony campaigns in order that the Army might apply those insights on today's battlefield. Particular attention will be paid to the possible threads of continuity with regard to force design and organizational considerations for each type of cavalry employed: light, medium, and heavy. Finally, this study will summarize the primary insights gained from both campaigns with regard to cavalry operations and will pay particular attention to any operational and tactical techniques that may be considered for future decisive, shaping, or sustaining operations.

For the sake of clarity and consistency, this study has listed the definitions of the levels of war and the three types of military operations discussed in the analysis portions of the chapters. The source for these definitions is the United States Army's current FM 3-0, *Operations*, dated June 2001.

The Levels of War

Strategic Level of War. The strategic level is that level at which a nation, often as one of a group of nations, determines national and multinational security objectives and guidance and develops and uses national resources to accomplish them.[1]

Operational Level of War. The operational level of war is the level at which campaigns and major operations are conducted and sustained to accomplish strategic objectives within theaters or areas of operations. It links the tactical employment of forces to strategic objectives.[2]

Tactical Level of War. Tactics is the employment of units in combat. It includes the ordered arrangement and maneuver of units in relation to each other, the terrain, and the enemy to translate potential combat power into victorious battles and engagements.[3]

The Types of Operations

Decisive Operations. Decisive operations are those that directly accomplish the task assigned by the higher headquarters. Decisive operations conclusively determine the outcome of major operations, battles, and engagements. There is only one decisive operation for any major operation, battle, or engagement for any given echelon. The decisive operation may include multiple actions conducted simultaneously throughout the area of operation. Commanders weight the decisive operation by economizing on combat power allocated to shaping operations.[4]

Shaping Operations. Shaping operations at any echelon create and preserve conditions for the success of the decisive operation. Shaping operations include lethal and nonlethal activities conducted throughout the area of operations. They support the decisive operation by affecting enemy capabilities and forces or by influencing the opposing commander's decisions. Shaping operations use all the elements of combat power to neutralize or reduce enemy capabilities. They may occur before, concurrently with, or after beginning of the decisive operation. They may involve any combination of forces and occur throughout the area of operation.[5]

<u>Sustaining Operations</u>. Sustaining operations are operations at any echelon that enable shaping and decisive operations by providing combat service support (CSS), rear area and base security, movement control, terrain management, and infrastructure development.

---

[1]Department of the Army, FM 3-0, *Operations* (Washington, DC: Government Printing Office, 2001), 2-2.

[2]Ibid.

[3]Ibid., 2-5.

[4]Ibid., 4-23.

[5]Ibid.

CHAPTER 2

NAPOLEON'S CAVALRY

The purpose of this chapter is to provide a foundation for understanding Napoleon's three types of cavalry. By understanding how they were each organized, equipped, and employed, it will be possible to gain a greater understanding as to how cavalry impacted Napoleon's battles and campaigns. This chapter will also set the conditions to better understand at which level of war and which type of military operation cavalry units made their greatest contribution. Any parallel insights to possible force structure considerations for today's forces will also be considered.

The Beginning

In order to understand the impact the French cavalry had on Napoleon's campaigns, one must first understand what Napoleon had to work with and how he planned to use it. When he rose to power in 1799 as First Consul, Napoleon inherited eighty-five regiments of cavalry.[1] While this may initially appear to be a substantial number of mounted units, his French forces did not compare to the quality of Prussian or Austrian cavalry. The eighty-five regiments were grouped into three different categories: thirty-eight light regiments, twenty medium regiments, and twenty-seven heavy regiments of cavalry, each type being employed in various manners. As Gunther Rothenberg described: "Convinced that it was not possible to fight anything but a defensive war without at least parity in cavalry, Napoleon made great efforts to turn this branch into a powerful striking force, capable of rupturing the enemy front, while retaining its ability for exploitation, pursuit and reconnaissance."[2] By the end of the

Napoleonic era, the quality of the French cavalry would be greatly improved compared to its modest beginnings.

Light Cavalry

Napoleon's light cavalry consisted of hussar, *chasseurs-a-cheval,* and lancer regiments, although the lancers were not formed until later. The Lancer's greatest increase in size occurred in 1811, just before Napoleon's impending invasion of Russia. Prior to the French Revolution, the hussar regiments consisted primarily of foreign soldiers and were based on the Hungarian light cavalry, from whence they derived their name. By 1800, however, the hussar units no longer relied on mercenaries and consisted mostly of French troops.[3]

The primary missions given the light cavalry were reconnaissance, screening, advance guard, and pursuit missions. They could also be subdivided into smaller-sized units for use as pickets and vedettes (mounted sentinels deployed forward of an outpost). As British historian Sir Charles Oman describes, the hussars were, "Intended to be the lightest of light cavalry, and were to find their proper sphere in raids and reconnaissance rather than in pitched battles."[4] Napoleon relied on his light cavalry to gain and maintain contact with the enemy and to screen his movements. A successful screen would deny the enemy valuable information with regard to the location, size, and composition of Napoleon's forces. The light cavalry was also employed as couriers and used to secure the French lines of communication. Along with reconnaissance, however, one of the most significant contributions the light cavalry made to Napoleon's campaigns was in the role of pursuit. Often it was the use of the light cavalry, pursuing a defeated and retreating enemy, which proved decisive in completing the destruction of the enemy force. A classic example of this was the French pursuit of the Prussian army following the battle of Jena in 1806. This is a classic example of what Napoleon meant when he said, "It is for the cavalry to follow up the victory and prevent the beaten enemy from rallying."[5]

Putting these light cavalry missions in the context of the current doctrinal definitions, the reconnaissance, screening, and advance guard missions of the light cavalry would be considered types of shaping operations. The use of light cavalry as couriers and to secure the lines of communication are examples of sustaining operations. The cavalry pursuit missions could be considered shaping operations, but most likely would be considered decisive operations.

Light cavalry units, particularly the hussars, were also known to have an extremely bold and audacious reputation. To further enhance their mystique, the hussars were the most flamboyantly dressed cavalry units of any. They based their uniforms on the Hungarian light horse units and wore Hungarian cut breeches, braided uniform shirts, and braided dolmans (jackets) often worn over one shoulder. All units wore shakos for headgear with the exception of the *Compagnies d'Elite* that wore fur busbies instead. They were lightly equipped with a heavy, curved saber for slashing, and carried one or two pistols, and a carbine.[6]

The other type of light cavalry unit Napoleon employed was the lancers. Although he did not have any organic lancer units when he assumed the throne, he did employ Polish volunteer units, the Lancers of the Vistula, who fought for the French during the Battle of Wagram. Napoleon was so impressed with the lancer's capabilities that he eventually stood up nine regiments of his own. The greatest increase in their numbers occurred 1811 during Napoleon's buildup of forces in anticipation of the impending war with Russia. It was during this time period that Napoleon converted six of his medium cavalry (dragoon) regiments into lancers.[7]

Typically, these units were armed with a lance, a saber, and a pistol. The lance was approximately nine feet long, one inch in diameter, made of a hard wood, such as ash or walnut, and weighed approximately seven pounds.[8] With its extended length, the lance also afforded its owner three distinct advantages over the saber. First, during cavalry on cavalry melees, the lance increased the shock effect on the opponent by being able to engage the enemy before he could effectively use his saber. Second, the lance proved superior to the saber when attacking the infantry squares. The infantry would typically form into squares to defend themselves against cavalry charges and relied on their bayonets once they had expended their rounds. Because of the lance's extended reach, lancer units were sometimes employed as the breach force unit to penetrate the infantry squares. This was especially true in the case of foul weather. During the 1813 battle of Dresden, heavy rains dampened the gunpowder, thus decreasing the chances for discharge. Consequently, the Austrian infantry formed in squares and was able to withstand initial French attempts at penetration. To overcome this, the French cavalry commander, Marshal Murat, effectively used his lancers as the breach force element to successfully penetrate the enemy line. He then followed through the penetration with his heavy cavalry, the *cuirassiers*, as the assault force to destroy the infantry squares. Last, the extended reach of the lance proved far more effective than the saber during pursuit missions when it was the cavalry's role to chase down and destroy enemy units attempting to escape. Although the lance did provide a significant advantage over the saber, its one drawback occurred during extremely close combat. Because of the extended length of the weapon, the lance often became too awkward and cumbersome for close-in fighting. The addition of the lancers to French cavalry organizations was one of the more-significant contributions Napoleon made to the mounted combat arm of decision.

Medium Cavalry

Napoleon's next category of cavalry was his medium cavalry, better known as dragoons, of which he inherited twenty regiments. The dragoons were equipped with a long straight sword (for thrusting), pistols, a dragoon musket (which was shorter than the infantry models), and a bayonet.[9] They typically wore brass helmets and tall boots, which were unsuited for dismounted action. Because of their mobility and increased firepower, as compared to other cavalry units, dragoons were used to seize key terrain for the main body or employed on the flanks with security force missions, all of which are examples of shaping operations using current doctrinal terms. Dragoons were also employed as battle cavalry for charges and were used extensively as mounted infantrymen in Spain.[10]

Napoleon found himself in the middle of an age-old debate of whether the dragoons were mounted infantrymen or cavalrymen with increased firepower. During the 30-Year War, dragoons were primarily mounted infantrymen. As Sir Charles Oman describes, "They were men with firearms who had been provided with horses in order that they might move rapidly, not light cavalry furnished with a musket for skirmishing purposes."[11]

By the eighteenth century, however, dragoons became more like cavalry and less like mounted infantry. For example, Frederick the Great employed his dragoons as cavalry with carbines or muskets. Because of their speed of mobility and firepower, Frederick's dragoons were expected to seize ground when infantry units were unavailable, and take charge of the skirmish line. Thus Frederick capitalized on the cavalry trait of mobility to shore up a potential weak point on the battlefield.

As Napoleon considered the force structure of his military at a junction, he turned the role of the dragoon back to that of mounted infantry. As such, he ordered the replacement of the knee-high boots with gaiters to aid in dismounted operations. Napoleon even went as far as planning to use dragoons as mounted infantrymen for his cross-channel invasion of England. One interesting side note, however, was that due to the lack of horses Napoleon's invasion plan called for two divisions of dismounted dragoons to utilize captured horses once in England. Fortunately for both sides, the invasion never occurred.[12]

One drawback of Napoleon's dragoons was that as they ceased to be effective cavalry, they degenerated into mediocre infantry as well. Because they had horses, they tended to stay mounted and their dismounted skills waned. But because they were also expected to perform as dismounted infantrymen, their performance with cavalry and maneuver skills suffered. Consequently, after 1807, Napoleon abandoned the idea of using dragoons as mounted infantry and decided to return the dragoons to their original role of medium cavalry.

After Napoleon seized Spain in 1808, twenty-four of the thirty dragoon regiments were transferred to the peninsula. This was where the majority of the dragoon regiments remained for the rest of Napoleon's reign. As a result, only six of Napoleon's thirty dragoon regiments were available in 1809 for the war with Austria, and only four dragoon regiments accompanied him into Russia in 1812. After the disastrous results in Russia, where he lost fourteen *cuirassier* regiments, Napoleon was forced to start pulling dragoon regiments from Spain and refit them as heavy cavalry units for the Leipzig Campaign of 1813.[13]

Heavy Cavalry

The final category of cavalry Napoleon inherited was the twenty-five regiments of heavy cavalry.[14] The heavy cavalry was broken down into two types, the *cuirassiers* and the *carabiniers a cheval*. These were the big men on big horses who were held in reserve exclusively for service in battle. Due to their large size and heavy armor, which increased their protection and survivability, the heavy cavalry was Napoleon's decisive combat arm that could deliver a devastating blow upon enemy units when properly employed. In context of current doctrine, the heavy cavalry would be kept almost exclusively for decisive operations.

Typically heavy cavalry charges were used in conjunction with the artillery. Following an artillery barrage, the heavy cavalry charged forward in mass in order to penetrate enemy lines and exploit any tactical success. Napoleon also used his heavy cavalry to counterattack any enemy cavalry assault.

In order to preserve the combat effectiveness of the heavy cavalry in battle, the tasks of courier duty, screening, reconnaissance, and pursuit typically fell to lighter cavalry units so that the heavy cavalry could be employed with maximum effectiveness at the critical time in battle. Napoleon was even quoted as saying, "Under no consideration shall cuirassiers be detailed as orderlies. This duty shall be done by lancers; even generals shall use lancers. The service of communications, escort, sharpshooters, shall be done by lancers."[15]

The *cuirassiers* were also uniquely equipped. Their name derived from the metal breastplate, *cuirass*, they wore. To further increase their survivability, Napoleon ordered that a back plate be added to the cuirass as well as equipping these units with steel helmets. The structural criteria for the breastplate was specified to be able to withstand one shot "at long range."[16] While the cuirass did not necessarily prove effective against musket fire at short range, it could withstand shots from pistols as well as attacks from lances, sabers and bayonets. For offensive weapons the heavy cavalry troopers were issued a longer straight sword for thrusting, two pistols, and either a musketoon or carbine "so they could deal with small bodies of enemy infantry in villages or defiles."[17]

The *carabiniers a cheval* were similarly equipped but did not wear armor, like the *cuirassiers* until 1809. Originally known as the horse grenadiers, they were fitted with carbines instead of pistols for the Danube Campaign of 1809. They did, however, have the reputation of being hand picked and, therefore, the favored force, sometimes referred to as royal pets. Needless to say they developed the attitudes to match.[18]

Although Napoleon's heavy cavalry had the reputation of being well equipped and provided for, they did have their drawbacks. With regard to cuirassiers, Napoleon once stated, "One result of having men of large stature, is the necessity of large horses, which doubles the expense and does not render the same service."[19] Because of the size requirements for the horses, only large breeds were accepted into the regiments. As a result, Napoleon's resource base was limited to Normandy and parts of Germany where large, powerful horses were bred. The necessity for large horses also increased the time required to produce another mount to replace one lost in battle. In conjunction with this limitation, theses horses were also vulnerable to severe weather and were not particularly well suited for winter campaigning where foraging became a challenge for the large quantities of food required. Another drawback to the heavy cavalry regiments was the cost required to produce and maintain them. During the early nineteenth century the price for horses was approximately 300 francs for a *cuirrasier's* horse, 200 francs for a dragoon's horse, and a horse for the light cavalry would cost around 100 francs. The horses for the officers and guards, being of the highest quality, could run as much as 500 francs or the equivalent of $800 U S dollars.[20] Consequently, this kept the number of heavy regiments down. These drawbacks became painfully clear following the Russian Campaign of 1812 where Napoleon lost fourteen *cuirassiers* regiments. As historian Hew Strachan states, "The loss of horses on the 1812 campaign so crippled the cavalry that it never fully recovered."[21] This had a direct impact on the results of Napoleon's campaign in Saxony during 1813.

Employment of Cavalry

Understanding the types of cavalry units Napoleon had is important, but it is only part of the issue. The other part is understanding how the cavalry was employed in battle. According to British historian David Chandler, Napoleon's tactical methods involved three phases during which the cavalry played a critical role in each.[22] The first phase was the movement to contact in which the light cavalry, performing reconnaissance missions forward of the advancing main body, would establish contact with the enemy forces. This would set the conditions for the advance guard to fix the enemy, phase two. The second phase began as the main body's advance guard began to engage the enemy. While this was taking place, the light cavalry would then position themselves off to a flank in order to establish a screen line that would conceal the maneuver force's positioning from the enemy, prior to the impending flank attack. The third phase involved the reinforcement of the advanced guard's fight as they engaged the enemy in a battle of attrition. Once the enemy was fixed, Napoleon would then launch a flanking attack to cut off the enemy's line of retreat and force him to extend and fight in two directions at once. It was then at this apex of the line, where the enemy was typically weakest, that Napoleon selected for his point for penetration. A massed artillery bombardment would devastate the weakened enemy line, and the heavy cavalry would be committed to penetrate the line and exploit the enemy. Once the artillery and heavy cavalry created the gap, the light cavalry would then be committed to follow through and begin the pursuit.

As Sir Charles Oman writes, "The main duty of Napoleon's cavalry then, was to make its weight felt in battle, urge pursuits to the extreme limit possible, and to screen the advance of the main columns, which it covered, on each road that they were using, at a moderate distance to the front."[23] To do this Napoleon kept his cavalry massed together as a "cavalry reserve" consisting primarily of dragoons, *cuirassiers*, and *carabiniers a cheval* to be committed at precisely the right time and place to exploit tactical success on the battlefield. This is how the cavalry contributed to massing the effects of combat power. The light cavalry then, was the force used for screening and pursuit missions. This cavalry reserve force typically stayed with the main body or striking force of the army so they would be ready to assist in the annihilation of the enemy force once brought to battle. Table 1 depicts the typical missions and capabilities of the three types of cavalry that existed in the early nineteenth century.

Type of Cavalry	Typical Missions	Attributes	Before Battle	During Battle	After Battle	Comments
Light Cavalry Hussars *Chasseurs-a-cheval* Lancers	Reconnaissance Screening Advance Guard Pursuit Courier Duty LOC Security Pickets Vedettes	Critical to Recon- Situational Understanding for the commander Important to C2 (Courier Duty) Pursuit – rendered destruction of enemy complete	Type of Operation Shaping Recon Impact of Operation Operational Tactical	Type of Operation Shaping Flank Security Courier Impact of Operation Tactical	Type of Operation Decisive Pursuit Impact of Operation Strategic Operational Tactical	Could contribute to operations in all three levels of war Greatest contribution was usually before or after the battle Least expensive force Shortest regeneration time
Medium Cavalry Dragoons	Seize Key Terrain Flank Security Hasty Attack Penetration (by exception)	Seized Key Terrain Gained Positional Advantage for maneuver forces Could perform hasty attacks and penetrations if required	Type of Operation Seize Key Terrain Impact of Operation Operational Tactical	Type of Operation Shaping Security Decisive If executing a hasty atk or penetration Impact of Operation Tactical	Type of Operation Decisive If conducting Pursuit Impact of Operation Strategic Operational Tactical	Most versatile force Capable of conducting light or heavy cavalry missions Back up force for the other two types of cavalry Greatest contribution was typically before or during battle Moderately expensive and regeneration time
Heavy Cavalry *Cuirassiers* *Carabiniers-a-cheval*	Penetration Deliberate Attack	Clearly a decisive element on the tactical battlefield	Type of Operation N/A Impact of Operation N/A	Type of Operation Decisive Delib Atk Penetration Impact of Operation Tactical	Type of Operation N/A Impact of Operation N/A	Least versatile Held exclusively for Decisive Operations (Tactical Level) Greatest contribution was during the battle Most expensive force Required longest time to regenerate losses

Table 1. Cavalry Missions and Capabilities of the Early Nineteenth Century.

As Napoleon wrote, "Cavalry charges are equally as good at the beginning, during, or at the end of a battle; they ought to be undertaken whenever they can be made against the flanks of infantry, especially when the latter is engaged in the front."[24] The technique used during the cavalry charges would be to charge forward as closely as possible in order to concentrate the massed effect on the enemy. Often the charges would be made in successive waves in order to achieve the full shock effect against the enemy.[25]

It was through this sequence for battle and execution of the charge and pursuit, that the French cavalry became a formidable European force. By 1807 the French heavy cavalry regiments had the reputation of being known as "the dread of Europe and the pride of France" while the light cavalry regiments were known for their "panache, daring, and gallantry."[26]

Napoleon's cavalry enjoyed their effectiveness on the battlefield until 1813 when two important events occurred. The first was that the French cavalry was unable to fully recover from the devastating Russian Campaign of 1812. The second was that Napoleon's enemies began adopting his methods of warfare and used them against him.

Future Force Structure Considerations

With this historical setting as a reference, it is now possible to draw a few insights with regard to today's force structure capabilities. Currently, U.S. Army mounted forces are suited for either extreme of the intensity scale of conflict. Consequently, the forces are either too heavy or too light. There exists no medium or "middle" capability.

The heavy forces, with the Abrams family of main battle tanks, are seventy-ton vehicles capable of conducting decisive operations over a wide variety of terrain. It is the modern day version of the *cuirassier*. However, just like the cuirassier, the main battle tanks cannot be the only type of mounted combatant vehicle for all battlefields. While it is well suited for decisive operations and some shaping operations, such as screening missions or advance guard, it is not suited for reconnaissance and would be considered a waste of combat power to relegate tanks to sustaining type operations.

Another member of the heavy force is the Bradley fighting vehicle (BFV) belonging to mechanized infantry units. The BFV, weighing over twenty-five tons, is designed to carry infantrymen into battle while also possessing some degree of lethality with their main gun (25 millimeter) and antitank missile launcher. Current mechanized infantry units could be described as the modern day version of the dragoon. They have the maneuverability of the main battle tank, but also have the dismounted forces that enhance their effectiveness at seizing terrain. Similar to Napoleon's dragoons, the mechanized infantry forces have the ability to conduct both shaping and decisive operations. They can perform reconnaissance missions, screening missions, and hasty attacks. Where they are lacking is in the ability to conduct some deliberate attacks or penetrations that would require breaching operations against antitank minefields. Another drawback to the BFV, just like the Abrams family of main battle tanks, is its weight. Because the BFV weighs over twenty-five tons and the Abrams over seventy tons, both are considered strategic mobility challenges and not as rapidly responsive as desired due to airlift limitations.

On the other end of the scale is the HMMWV, typically a lightly armored wheeled vehicle capable of conducting selected shaping and sustaining operations such as reconnaissance and securing lines of communication. It cannot, however, conduct decisive operations in a high intensity conflict, and also lacks the survivability and lethality to successfully conduct screening or pursuit missions.

In order to bridge this capabilities gap, the U.S. Army is developing the IBCT with a wheeled, lightly armored vehicle (LAV) of medium weight, approximately twenty tons or less. In several ways this parallels Napoleon's concept of the dragoons with one exception, the LAV is expressly an infantry-centric platform. Therefore, the IBCT is not faced with the dilemma of the dragoon--are the troopers mounted infantry or dismounted cavalry? They are clearly mounted infantrymen. In this aspect, the LAV is well suited to conduct sustaining operations, such as securing lines of communication, and some shaping operations, such as securing key terrain and reconnaissance. The IBCT would, however, find the missions of screening or pursuing mounted enemy forces challenging since the LAV is lacking in both lethality and survivability. Granted, it is considerably more lethal and survivable than a dismounted infantryman, but against a heavily armored mounted opponent it is not well suited.

These considerations for the current U.S. Army force structure capability should be kept in mind throughout this work and will be revisited in the final analysis chapter.

---

[1] John R. Elting, *Swords Around a Throne* (New York: The Free Press, 1988), 229.

[2] Gunther E. Rothenberg, *The Art of Warfare in the Age of Napoleon* (Bloomington, Indiana: Indiana University Press, 1978), 141.

[3] Sir Charles Oman, *Studies in the Napoleonic Wars* (London: Methuen and Company, Ltd., 1914), 243.

[4] Ibid., 243.

[5] LTC Ernest Picard, "Maxims and Opinions of Napoleon on the Use of Cavalry," *Journal of the U.S. Cavalry Association* 24 (July 1913 to May 1914): 1002.

[6] Elting, 239.

[7] Elting., 235.

[8] Elting, 242.

[9] Elting, 238.

[10] Rothenberg, 141.

[11] Oman, 235.

[12] Elting, 236.

[13] Oman, 241.

[14] Elting, 229.

[15] Picard, 997

[16] Elting, 230.

[17] Elting, 233.

[18] Elting, 234.

[19] Picard, 996

[20] Z. Grbasic and V. Vuksic, *The History of Cavalry* (New York: Facts on File, 1989), 133.

[21] Hew Strachen, *European Armies and the Conduct of War* (London: George Allen and Unwin, 1983), 54.

[22] David G. Chandler, *Campaigns of Napoleon* (New York: Scribner, 1966), 187.

[23] Oman, 253.

[24] Picard, 1001.

[25] Elting, 540.

[26] Rothenberg, 142.

CHAPTER 3

THE JENA CAMPAIGN

Along with Napoleon's 1805 victory at Austerlitz, the Jena Campaign of 1806 ranks among the greatest military achievements of his entire reign. This campaign demonstrated Napoleon's unquestioned genius as a strategic planner, mastery of the operational art of maneuver, and flexibility as a tactician. Inherent to the success of this campaign was the French cavalry. The cavalry not only performed admirably by setting the conditions for and during battle, but the cavalry also succeeded in executing one of the greatest pursuits in history--the result of which was the complete destruction of the Prussian army.

In order to fully comprehend the French cavalry's contribution to this campaign, this chapter will first discuss the historical background and the conditions that led to the road to war. A chronological description of the campaign will then be reviewed leading to an analysis of cavalry's impact in determining the outcome of the campaign. For the analysis portion, each type of cavalry (light, medium, and heavy) will be examined within the context of the three levels of war (strategic, operational, and tactical). The analysis will also examine the cavalry's employment in the three types of military operations (decisive, shaping, and sustaining). It is also the intention of this study to gain battlefield insights from this campaign that might be applied to today's mounted operations. A depiction of the French order of battle is depicted in Table 2.

The Road to War

With Austria's defeat at the battle of Austerlitz and its signing of the Treaty of Pressburg in December 1805, Napoleon succeeded in changing the political landscape of

Europe by gaining control of western and southern Germany.[1] By July 1806 he established the Confederation of the Rhine that included most German states with the exception of Prussia.

Prussia, which had allied itself with France after the battle of Austerlitz in December 1805, had done so on the condition that it would gain control of Hanover. This alliance was jeopardized the following summer when the Prussian ruler, King Frederick William III, heard rumors that Napoleon was offering Hanover as peace token to Britain. The threat of the loss of Hanover, compounded by the losses suffered in the Confederation of the Rhine, proved to be an intolerable diplomatic situation for Prussia.[2]

In July 1806, Prussia took another diplomatic step pushing the nation closer to war. She allied herself with Russia, who was still at war with France. In August, the Prussians held a council of war at Potsdam resulting in the decision to mobilize the Prussian army. This mobilization included an army of approximately 250,000 men, of which about 145,000 men could be deployed as a field army: General Wurttemberg with 15,000men; General Ruchel with 28,500 men; The Duke of Brunswick with 60,750 men; and General Hohenlohe with 42,000 men. Along with this, the Russians agreed to reinforce the Prussians with 120,000 men in the form of two armies of 60,000 each. The smaller German states of Brunswick, Hesse-Cassel, and Saxony also allied themselves with Prussia.[3] Not desiring to resume hostilities, Napoleon proposed compromise. He informed the Prussian ambassador that French soldiers would withdraw across the Rhine provided that Russian soldiers would be sent home.[4]

Meanwhile, convinced that Prussia once again had an army as powerful as the one commanded by Frederick the Great, King Frederick William III decided to launch his attack against France without waiting for the arrival of the Russian armies. On 12

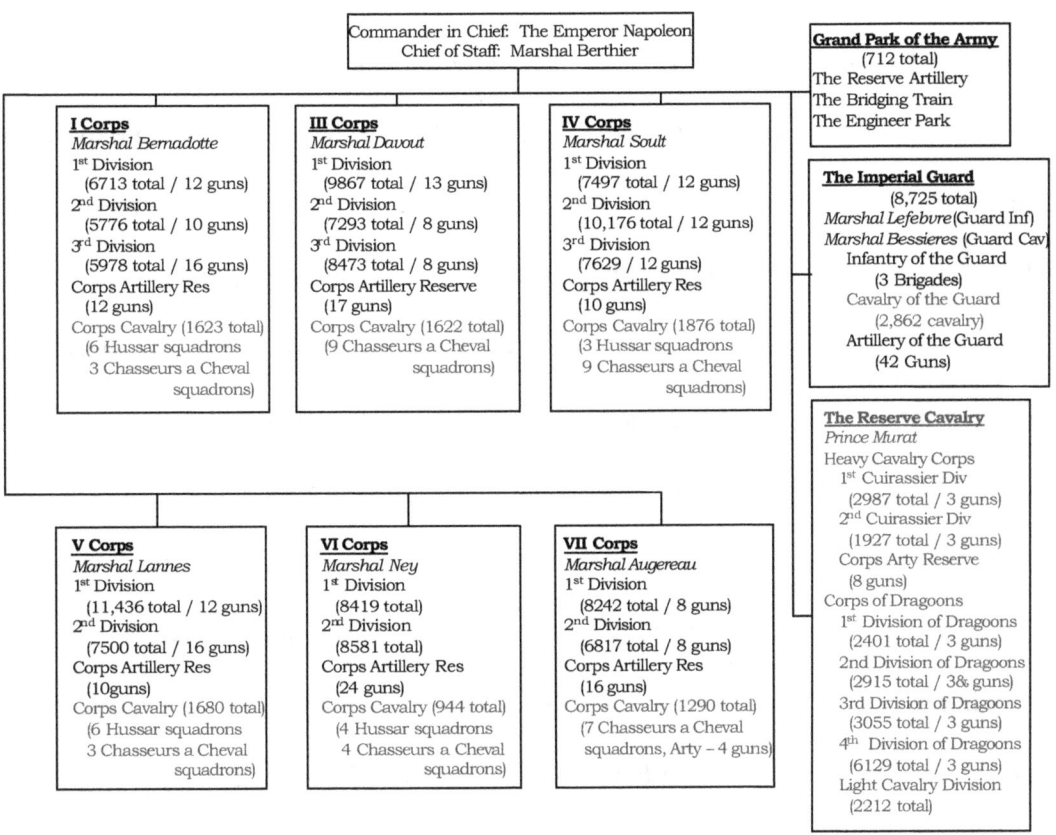

Table 2. French order of battle for the Jena campaign as of October 1806. Source: David G. Chandler, *Jena 1806, Napoleon Destroys Prussia* (Oxford: Osprey Publishing Ltd., 1993), 34-37.

September the Prussian forces marched into Saxony, this action alone could have been interpreted as an act of war against France.[5] Prussian forces continued massing with the intent of launching a surprise attack towards the Main River in order to cut off the French lines of communication (LOC). By 8 October Prussian forces stretched along a sixty mile front extending from Eisenach to Jena.[6]

During the posturing of Prussian forces, Napoleon was kept informed of the Prussian and Russian intentions by his ambassadors and *attaches*. They informed him of three critical pieces of information. One, they confirmed the Prussians planned to attack the French forces in Germany. Two, they reported that the Russians would not be able to mobilize their army for several weeks still. And three, Napoleon was able to determine that his forces, numbering approximately 200,000 men, would vastly outnumber the Prussians, if they could be massed against the Prussian alone.[7]

It was Napoleon's intent to defeat the Prussians before their new Russian allies could reinforce them. He decided to position his forces in northern Bavaria and strike north, through the Thuringerwald Mountains with the purpose of cutting off the Prussian LOC that extended from Berlin southwest to Erfurt. This was particularly appealing to Napoleon since he was aware of the vulnerability that Prussian supply depots and slow moving supply convoys presented.[8] If successful, this plan would also result in Napoleon positioning his forces between the Prussians and their Russian allies to the east.[9]

Based on his intelligence reports, Napoleon issued over 102 orders and letters to his staff on the 18 and 19 September preparing his Grande Armee for the impending war with Prussia.[10] In issuing his "General Dispositions for the Assembly of the Grand Army" Napoleon directed his forces be arrayed in the following manner: Augereua's VII Corps was to be positioned in the vicinity of Frankfurt; Lefebvre's V Corps to the vicinity of Konigshofen; Davout's III Corps and Bernadotte's I Corps to the vicinity of Bamberg; Ney's VI Corps to the vicinity of Ansbach; and Soult's IV Corps to the vicinity of Amberg. The Imperial Guard and Kellerman's reserve corps would march from Paris to Mainz, and Berthier was to relocate Napoleon's headquarters to the vicinity of Wurzburg. Murat's reserve cavalry corps was directed to collect the divisions of the cavalry reserve and position their forces ranging from Schweinfurt to Kronach.[11]

To ensure operational security, Napoleon forbade French forces from crossing into the Prussian and Saxon frontiers. French interior lines, remaining protected by the Main River as well as the densely wooded and mountainous terrain of the Thuringerwald and Frankenwald regions, would further help the French maintain their element of surprise.[1] Figure 2 depicts the situation as of 6 October 1806.

In a note to his brother, Louis, the King of Holland, dated 30 September, Napoleon described his intent for the impending campaign,

> My intention is to concentrate all of my forces on the extreme right, leaving the entire area between the Rhine and Bamberg unoccupied so as to permit the concentration of 200,000 men on the same battlefield. Should the enemy push forces between Mainz and Bamberg, it will not worry me for my line of communications is based on Forcheim, a small fortress, and thence on Wurzburg . . . The exact nature of the events that may occur is incalculable, for the enemy who believes my left is on the Rhine and my right in Bohemia, and thinks that my line of operations runs parallel to my battlefront, may see great advantage in turning my left, and in that case I shall throw him into the Rhine.[2]

On 2 October 1806, King Frederick William III, issued his ultimatum to Napoleon demanding French soldiers be removed from German soil. The suspense for a reply was 8 October. Ironically, the ultimatum did not reach Napoleon until 7 October, since he had been enroute to Wurzburg on the day it was issued. When Napoleon finally received the Prussian ultimatum for war, he responded by saying, "We have been challenged to the field of honor. No Frenchman ever missed such an appointment. I shall be in Saxony tomorrow."[3]

Figure 1. Jena Campaign. Situation 6 October 1806. Source: United States Military Academy History Department Maps. Available from http://www.dean.usma.edu/history/dhistorymaps/images/Napoleon/nap26.jpg

The Movement to Contact

Napoleon realized that the elements of success for this campaign would lie in the principles of surprise, speed and mass. He had a general idea of where the main enemy forces were located but was not entirely sure. He was also fairly confident that the enemy still believed his main force was further west than it actually was. As long as the enemy remained in the dark, his penetration to the northeast through the Thuringerwald Mountains would enable the French to maintain the initiative. Speed was of essence, however, and Napoleon couldn't allow his forces to get bogged down through a mountain pass, or choke point, that would facilitate the foreboding reinforcement of Russian troops. Napoleon also needed to be able to mass his forces at the decisive point and time on the battlefield to ensure victory over the Prussian forces. To achieve and maintain the momentum he devise a clever movement formation called the battalion box, or in French *"Le Bataillon Carre." Le Bataillon Carre* was in essence a massive wedge formation that had concentrated front of approximately thirty-eight square miles. Its depth could be covered in just two days, and the entire force of the Grande Armee could concentrate on any point in just two days as well. Napoleon described the reason for this formation in a note to Marshal Soult dated 5 October. Napoleon wrote, "You will understand, that with this enormous superiority of numbers, concentrated on such a small space, I have no intention to leave anything to chance, but to attack the enemy, wherever he wants to make a stand, with double the strength he can dispose of."[4]

Through careful map reconnaissance and reports from his engineer officers who had been dispatched to carefully examine the roads leading from Bamberg to Berlin, Napoleon selected three separate routes on which to send his *Grande Armee* through the

Thuringerwald.[5] Lannes and Augerau would advance along the western route running from Colburg to Saalfeld. Bernadotte, Murat, and the Imperial Guard would travel along the center route from Kronach to Schleitz; and Soult and Ney would advance along the eastern route extending from Hoff to Plauen.[6] Light cavalry squadrons headed each of the three columns with the mission of conducting route reconnaissance and establishing contact with the enemy.[7] Napoleon was still not certain of the location of the enemy's main body, but he did anticipate Soult's IV Corps, along the eastern route, making contact with the enemy first.[8]

The first contact with the enemy occurred the morning of 9 October, when forces along the center route, belonging to Bernadotte and Murat encountered 6,000 Prussians and 3,000 Saxons under the command of Generalmajor Tauenzien in the town of Schleiz. The French attacked with a force consisting of two light cavalry brigades, two divisions of dragoons, and infantrymen from Bernadotte's I Corps. The attack was successful and the northeast road to Gera was quickly opened.[9]

On the morning of 10 October, Lannes' forces made contact with a Prussian vanguard near the town of Saalfeld. Based on this contact report, Napoleon estimated that his rightmost column probably had no one remaining in front of it.[10] Napoleon quickly dispatched orders to Murat with the mission of conducting a reconnaissance in the vicinities of Auma and Saalfeld in order to determine the exact location of the enemy's main body. Napoleon believed that the Prussian main body was somewhere between himself, in the center column, and his western column under Lannes. A dispatch from Soult, who traveled in the eastern column, "rendered certain that the Prussian army intended to concentrate near Gera."[11] Napoleon was now confident that his forces could successfully reach Gera before the Prussians. In doing so, the French would succeed in cutting off not only the Prussian LOC, but also their escape route to Berlin. By the evening of 10 October, the French had captured Saalfeld and Napoleon ordered all columns to converge vicinity the town of Gera, with Lannes continuing his advance toward Neustadt. Figure 2 depicts the situation as of noon on 10 October 1806.

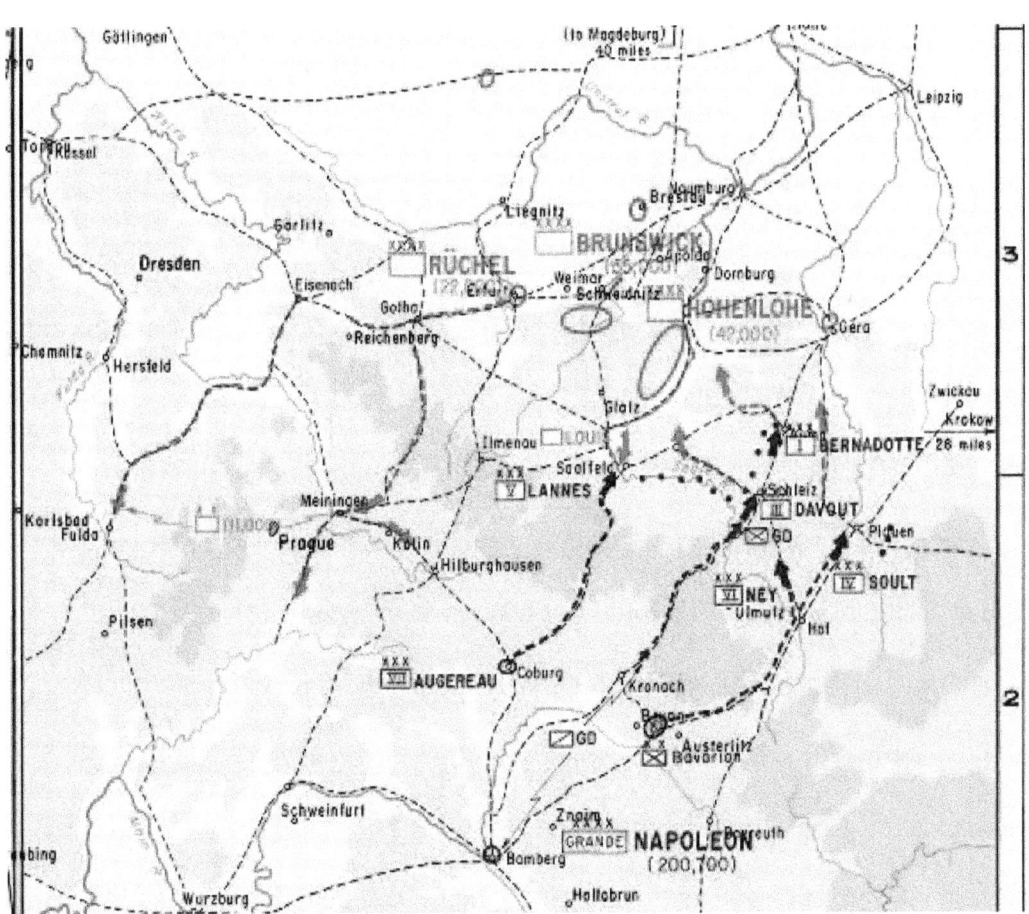

Figure 2. Jena Campaign. Situation Noon 10 October 1806. Source: United States Military Academy History Department Maps. Available from http://www.dean.usma.edu/history/dhistorymaps/images/Napoleon/nap27.jpg

By this time the Prussians, under command of Hohenlohe had fallen back upon the town of Kahla. They were clearly concerned about keeping open their LOC's along the Elbe River. Consequently, Prussian forces continued their retreat toward the town of Blankenhein, with the intent of regrouping with the Prussian army at Weimar.[12]

On 11 October, the French center column arrived at Gera, but found no enemy. The Prussians were still further west than Napoleon had anticipated. He therefore ordered his eastern column to reposition itself toward the center. Napoleon also ordered Murat's cavalry to continue their reconnaissance missions towards the towns of Zeitz and Jena to the west. Their instructions were to collect information on the enemy, cut off the Prussian way of retreat, capture any available Prussian supply stores, and to spread panic.[13] Napoleon was intent on blocking the enemy's line of retreat across the Elbe, but had still not located the enemy's main body.[14]

By now, the Prussians had concentrated their three forces in the following manner: Hohenlohe's force was located vicinity Jena; the main army under Brunswick was located at Weimar; and Ruchel's force was a Weimar as well.

On 12 October, Napoleon ordered his forces to shift their direction of travel from the north to the northwest. The French advance guard was to cross the Saale River at three sites: Lannes' V Corps and Augereaus' VII Corps at Jena, and Davout's III Corps and Bernadotte's I Corps at Naumberg and Zeitz. The French main body would remain in the vicinity of Gera with Soult's IV Corps and the Imperial Guard. Ney's VI Corps would be at the vicinity of Mittel Polnitz along with the heavy cavalry reserve. The light cavalry would continue its reconnaissance towards Leipzig. Napoleon's calculations still led him to believe that an encircling movement was possible with the likely enemy contact occurring on 15 or 16 October in the vicinity of the Saale River.[15]

A dispatch from Davout to Napoleon's headquarters, on 12 October, described the French efforts to locate the Prussian army and confirm whether or not the Prussian LOC, running from Naumburg to Leipzig, was secured. Davout wrote, "All the reports from the deserters, of prisoners and of the people of the country unite to proclaim that the Prussian Army is to be found at Erfurt, Weimar and environs. It is certain that the King [of Prussia] arrived at Weimar yesterday; I am assured that there are no troops between Leipzig and Naumbourg."[16]

This same day, Lasalles's brigades of the 5th and 7th Hussars maneuvered towards the town of Moelsen. Their task was to conduct a reconnaissance around the towns and vicinities of Pegau, Leipzig, Weissenfels, and Naumburg with the strict orders of not entering into the towns themselves. Their purpose was to spread terror in the enemy's rear with the message that, "The Emperor was coming to cut off their retreat to Berlin."[17] They were also to capture enemy convoys, and collect any information on the enemy. By 4 P.M. a squadron of the 7th Hussars had arrived at the town of Weissenfels and by dusk a squadron of the 5th Hussars was near Moelsen. At Moelsen, the 5th Hussars ran into contact with fifty Saxon dragoons but successfully managed to drive them into the Wohlitz Ravine.[18] By 9 P.M., the Mathis Squadron of the 5th Hussars, near Pegau, reported, "The panic-stricken inhabitants had expected the King of Prussia to be victorious! In a flash the streets were empty and not a soldier could be seen."[19]

Another element of the 5th Hussars, of company size led by Captain Pire, maneuvered towards Leipzig. By 10 P.M., his advance guard managed to successfully capture the guard post with its sentries. From the Saxon guards, the French discovered that two enemy battalions had departed Leipzig earlier that day headed for Dresden, and there were only fifty grenadiers and two small companies remaining in the town as a garrison force. With regard to Pegau, the French were told that only two sentries lightly guarded the town, but there were several civilians in town due to a fair. The end result was the fact that there were no substantial enemy forces in the vicinity of Leipzig.[20] Figure 3 depicts the situation as of 6 P.M., 12 October 1806.

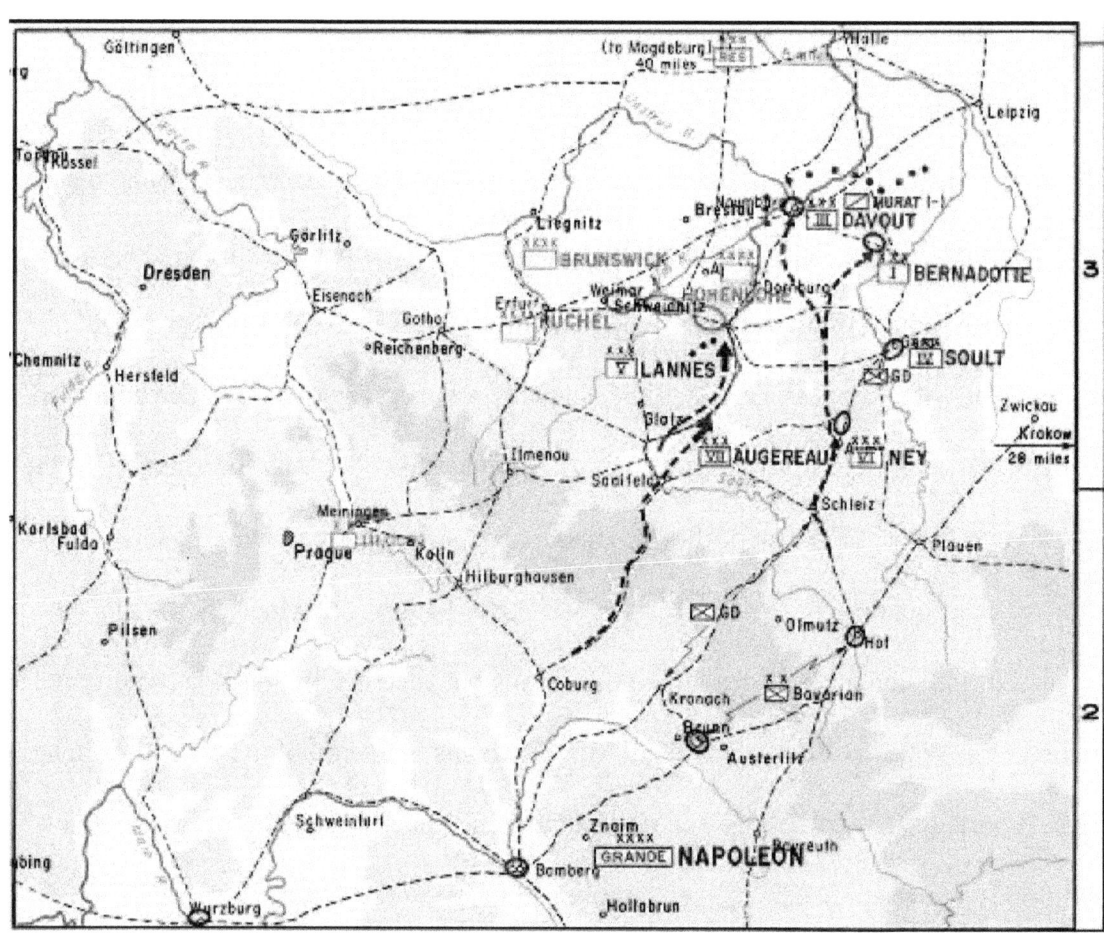

Figure 3. Jena Campaign. Situation 6 P.M., 12 October 1806. Source: United States Military Academy History Department Maps. Available from http://www.dean.usma.edu/history/dhistorymaps/images/Napoleon/nap28.jpg.

Finally, on the night of 12 October, Lannes V Corps made contact with Prussian outposts near the town of Jena. In a dispatch to Murat, written on the morning of 13 October, Napoleon wrote, "At last the veil is rent, and the enemy is beginning to retreat toward Magdeburg."[21] Napoleon now began to converge his forces on Jena for the decisive battle. His intent was to maneuver the entire French army to the west and engage the enemy. If the Prussians held their ground, he would attack them. If they were on the move he would halt them and compel them to fight.[22]

Napoleon arrived at Jena in the late afternoon of 13 October. He quickly conducted a personal reconnaissance of the high ground to the northwest of Jena called Landgrafenberg. From these heights Napoleon identified what he estimated to between 40,000 and 50,000 Prussians and realized the battle would the next day.[23] Napoleon saw his operation unfolding in two phases: the first was to secure the heights; the second was to destroy the Prussian army.[24]

Napoleon then ordered Lannes to move his entire V Corps to the high ground of the Landgrafenberg in order to secure the crossing sites along the Saale River for the converging French forces. He also sent orders to Lefebvre, Soult, Ney, and Augereau, along with 40 artillery pieces, to converge their forces on Jena.[25] Murat and Bernadotte were ordered to Dornburg, located six miles north of Jena. Davout was ordered to Naumburg with the orders to halt the Prussians. The heavy cavalry and dragoons from Klein, Soult and Ney were to position themselves in the vicinity of Roda.[26] Napoleon's main concern was that the Prussians would push the French off the heights before reinforcements could arrive.

Because of this, Napoleon directed his engineers improve and widen the road leading from Jena up to the heights of the Landgrafenberg. In an account from one of Augereau's staff officers, Baron Marbot, Napoleon, "Sent at once for 4,000 pioneering tools from the wagons of the engineers and artillery, and ordered that every battalion should work in turn for an hour at widening and leveling the path, and that as each finished its task it should go and form up silently on the Landgrafenberg while another took its place."[27] By sunrise, the French had managed to place forty-two artillery pieces and 25,000 men on the summit. Much to Napoleon's surprise the Prussians did not attack, but instead remained in their camps. Brunswick ordered his main army to begin their retreat from Weimar towards Auerstadt, with Hohenlohe covering his eastern flank. By the time the battle began, the Prussians would face 46,000 French troops and seventy guns massed in the vicinity of the Landgrafenberg.

Napoleon's cavalry played a critical role in setting the conditions for battle. They performed an effective advance guard through the Thuringerwald Mountains in front of the *Bataillon Carre,* which enabled Napoleon's main body to advance unmolested. Aggressive patrolling by light cavalry forces in the vicinity of Leipzig confirmed where the enemy was or was not located, and that the Prussian lines of communication remained unsecured to the east. Effective cavalry reconnaissance reports also provided Napoleon with important battlefield intelligence. This allowed him to accurately adjust his plans and position his forces so he could ultimately achieve positional advantage over his opponents. The cavalry clearly contributed to the operational success of the campaign. Figure 4 depicts the situation as of midnight 13 October 1806.

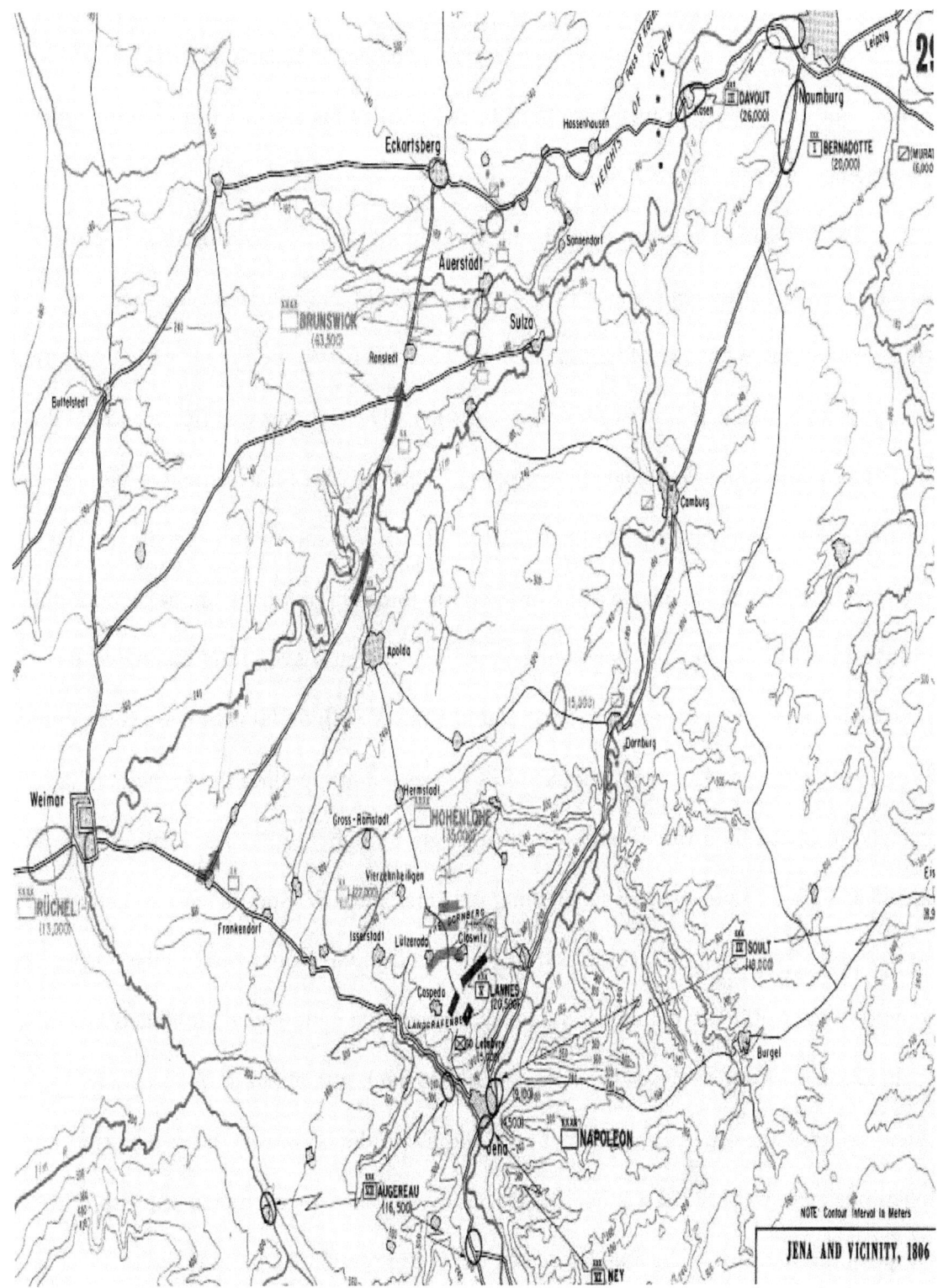

Figure 4. Battles of Jena and Auerstadt. Situation Midnight, 13 October 1806. Source: United States Military Academy History Department Maps. Available from http://www.dean.usma.edu/history/dhistorymaps/images/Napoleon/nap29.jpg.

The Battle of Jena

Historian David G. Chandler has described the battle of Jena in four distinct phases. For the purpose of this discussion we will utilize his construct in order to follow the sequence of events more clearly.

The first phase began at 6 A.M. and lasted until 10 A.M. During this time the French repelled Tauenzien's attack by his advance guard and enlarged the French bridgehead on the west side of the Saale River. Lannes began the French attack with two of his divisions supported by twenty-eight cannons. The fighting was fierce and the thick fog, still covering the ground, added confusion. The French captured the two villages of Closwitz and Cospeda and the Prussian counter attack was successfully repelled. The Prussian commander, Tauenzien, then ordered his forces to pull back and regroup to the northwest near the village of Vierzehnheiligen. Tauenzien successfully countered the French advance by launching his second counter attack with 5,000 men toward the center of the French forces. This counterattack was successful at repelling the French and also managed to divide their forces in two. Surprisingly though, the Prussians failed to press the attack. While Tauenzien was launching his counterattack, Soult's lead division made contact with forces under the command of Prussian General Holtendorff to the north. To the west, Augereau's second division advanced along the valley near Muhlbach towards the direction of Weimar. Fearing that his southern flank was about to be turned, Tauenzien gave the order for his forces to withdraw to the northwest in the direction of Hohenlohe's main forces. Thus by 10 A.M., Napoleon had succeeded in expanding the bridgehead to the west of Jena and securing the high ground of Landgrafenberg. Napoleon now had enough terrain to employ the rest of the Grande Armee and ordered his forces to halt in the center and south, while French forces in the north joined up.[28]

Phase two of the battle took place to the north of the Landgrafenberg with Soult's forces engaging Holtzendorff's 5,000 men. Prussian forces launched an attack against St. Hilaire's division with the main body in an echeloned line and the cavalry and twenty-two artillery pieces positioned on the flank. Fortunately for the French, the reverse slope of the terrain protected St. Hillaire's division. St. Hillaire then launched a counterattack against Holtzendorff's left flank utilizing a blind spot in the terrain. As a result, the Prussians began withdrawing across a stream near Nerkwitz. A screening force of Prussian cavalry and light infantry initially proved successful in covering this withdrawal. However, Soult's cavalry pressed the attack and succeeded in almost completely destroying one of the retreating Prussian columns. As Chandler records, they succeeded in capturing, "Four hundred prisoners, six guns and two colors."[29] Although Holtzendorff attempted to rally his men near the town of Nerkwitz, the French succeeded in launching a frontal cavalry attack against the Prussian forces. This proved too much for the Prussians who began abandoning their guns and retreated to the northwest towards the town of Apolda.[30]

This phase of the battle demonstrates two aspects of the cavalry that proved vital to the French success: the ability to pursue retreating forces, and the ability to counterattack enemy resistance. By pressing the attack and pursuing the retreating Prussians, the cavalry helped turn a retreat into a rout, resulting in the capture of over four hundred prisoners. By counterattacking the Prussians near the town of Nerkwitz, the cavalry succeeded in forcing the Prussian attack to culminate and ultimately forced them to retreat from the field of battle. These examples demonstrate the cavalry's ability to help seize the initiative and exploit success on the battlefield.

With the arrival of Tauenzien's forces, which had been forced off of the Landengrafenberg, Prince Hohenlohe took immediate action to regain the initiative. He ordered Ruchel's forces to advance from Weimar to reinforce Hohenlohe's main body in the east, while Tauenzien's forces were ordered to regroup and refit for action. Hohenlohe then ordered the majority of his forces to advance to the east in order to force the French off the Landengrafenberg plateau.[31] Figure 5 depicts the situation as of 10 A.M., 14 October 1806.

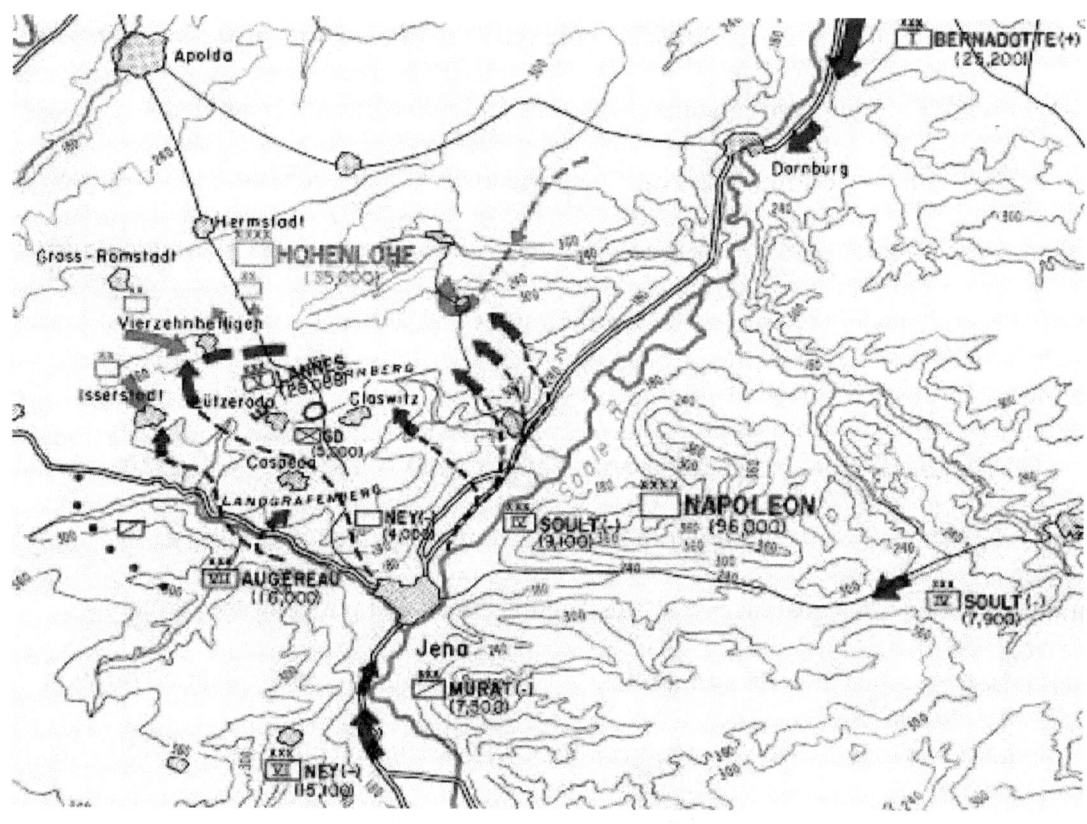

Figure 5. Battle of Jena. Situation 10 A.M., 14 October 1806. Source: United States Military Academy History Department Maps. Available from http://www.dean.usma.edu/history/dhistorymaps/images/Napoleon/nap30.jpg.

By 11 A.M. the Prussians had managed to emplace eleven divisions opposite Lannes' V Corps. However, the third phase of the battled occurred near the village of Vierzehnheiligen when Ney, without orders, directed two regiments of light cavalry and five infantry battalions to attack.[32] Although Ney's impetuous attack initially proved successful, he soon attacked beyond the range of the supporting French units on his flanks, Lannes to his right and Augereau to his left. The Prussian cavalry counterattacked and forced Ney's units to form into squares in order to withstand the assault. Napoleon realized what was happening and ordered his cavalry reserve, which only consisted of two cavalry regiments at that time, into the attack. He then ordered both Lannes and Augereau to continue their advance in order to link up with Ney's isolated forces. The cavalry counterattack and the advances by Lannes and Augereau succeeded in relieving the pressure on Ney, but Ney's impetuous behavior had cost the French unnecessary casualties.[33]

The fourth phase of the battle came when Hohenlohe reversed his previous decision and halted the Prussian advance. Rather than forcing the French off the plateau with the Prussian forces at hand, Hohenlohe decided to halt and wait for Ruchel's reinforcements to arrive from Weimar. As Chandler records, "Now followed one of the most extraordinary and pitiful incidents in military history. This line of magnificent infantry, some 20,000 strong, stood out in the open for two whole hours while exposed to the merciless case and skirmishing fire of the French, who, behind the garden walls, offered no mark at all for their return fire. In places the fronts of the companies were only marked by individual files still loading and firing, while all their comrades lay dead and dying around them."[34]

The French, with 54,000 men, pressed the attack against the stalled Prussians and managed to separate three Saxon brigades from the Prussian main body. By 1 P.M., Hohenlohe had expended all of his reserve forces in an effort to fill the gaps in his line. Unfortunately for the Prussians, Ruchel's forces would not possibly arrive until 3 P.M. at the earliest.

During all of this, Napoleon continued to mass his forces across the bridgehead and had assembled a reserve force of 42,000 men by half-past noon. This force consisted of Murat's cavalry and the main elements of Soult's IV Corps and Ney's VI Corps.[35] Napoleon ordered a general attack against the Prussian line, but did not have all of his forces in their prescribed positions until 1 P.M. Once all the French forces were set, Napoleon ordered his center force, consisting of Lannes V Corps and Ney's VI Corps, to advance. The French main body had now been committed.

Although a few Prussian artillery and cavalry units attempted to halt the oncoming French onslaught, their efforts proved in vain and the French succeeded in penetrating the Prussian line. Seeing this, Hohenlohe, ordered a general retreat of the Prussian forces toward the villages of Gross and Klein-Romstedt to the northwest. The Prussians initially succeeded in withdrawing some of their units in contact, but the situation grew considerably worse for the Prussians when Lannes brought his artillery forward and began engaging the retreating columns with harassing fires. Shortly after this, Murat launched his cavalry against the Prussians and transformed the orderly retreat into a rout. To the west, 15,000 of Ruchel's troops from Weimar finally appeared on the battlefield. However, instead of establishing a defensive line to cover the withdrawal of the retreating Prussian forces, Ruchel's forces began a retreat of their own. Napoleon pressed the fight and ordered more French cannons forward. Murat then committed his *cuirassiers* against Ruchel's forces and the Prussian rout began for Ruchel's forces as well. Murat continued his pursuit and by 4 P.M. rode triumphantly into the town of Weimar.[36] Napoleon returned to his headquarters that evening confident he had defeated the main forces of the Prussians. In the end, the battle of Jena resulted in approximately 25,000 Prussian losses compared to only 5,000 French.[37] Figure 6 depicts the situation as of 2 P.M., 14 October 1806.

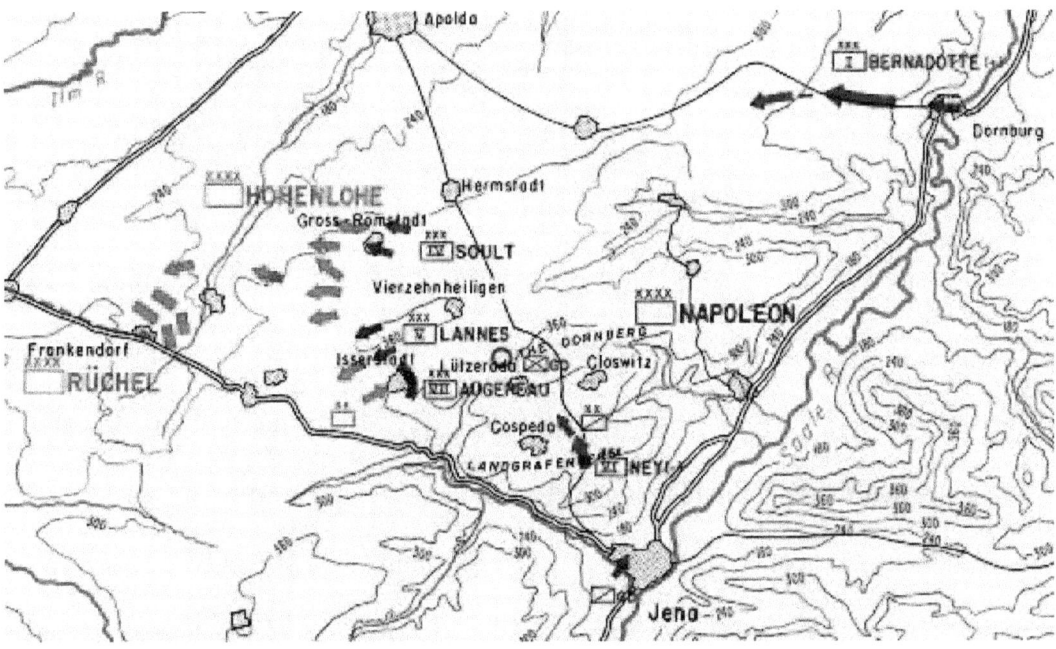

Figure 6. Battle of Jena. Situation 2 P.M., 14 October 1806. Source: United States Military Academy History Department Maps. Available from (http://www.dean.usma.edu/history/dhistorymaps/images/Napoleon/nap31.jpg)

Much of Napoleon's success at Jena was directly related to his cavalry's superb performance. Prior to the battle, his cavalry not only protected his forces by performing an effective advance guard, but they also collected vital information through thorough reconnaissance efforts. This allowed Napoleon to gain an accurate picture of the enemy and position the French forces to take advantage of the terrain. But cavalry's contributions did not stop there. During the battle, the cavalry continued to have a decisive impact. Not only did they effectively repulse enemy attacks, they also exploited aggressive pursuits against retreating forces that resulted in hundreds of enemy soldiers being captured. Following the battle, Murat was even able to capture the enemy town of Weimar, due to his aggressive cavalry exploitations. The cavalry clearly played a decisive role in helping Napoleon achieve his victory at Jena.

The Battle of Auerstadt

On 14 October at 4 A.M., Davout received his orders from Napoleon and began preparations to move his III Corps from Naumburg southwest to Apolda. Davout forwarded a copy of his orders to Bernadotte so that he and his I Corps maintained situational awareness of III Corps' activities. Since the fog was thick that morning, Davout began his movement under obscuration and without knowing the exact location of the Prussian forces. By 7 A.M., Davout's III Corps was passing through the village of Hassenhaussen when the lead French elements made contact with a Prussian cavalry screening force near the hamlet of Poppel. The French immediately opened fire and succeeded in forcing the Prussian screening force back across the Liss Bach stream. At this point the lead French forces halted in order for the rest of their element to join up. Meanwhile, the Prussian main body continued its movement to the northeast and by 8 A.M., the French lead element faced nine Prussian battalions, twenty-four cannons, and sixteen squadrons of cavalry.[38]

The main battle began with an exchange of fire from the skirmishers on both sides. However, it wasn't long until things changed. The Prussian cavalry, led by Blucher, launched a premature attack against the French before the Prussian infantry was in position. The French infantry squares had no problem in repulsing the uncoordinated attack. The Prussians continued to delay the launching of their main attack in order to have all of their forces in position. This delay only provided the French with the much needed time they required to position the rest of their forces. As Chandler describes, "Davout realized that the enemy's aim was to attack his right flank in order to keep the main road to Freiburg open, and so he pulled Gudin's division out of Hassenhaussen and redeployed it to the north of the village, less one regiment left to the south."[39] Figure 7 depicts the situation as of 10 A.M., 14 October 1806.

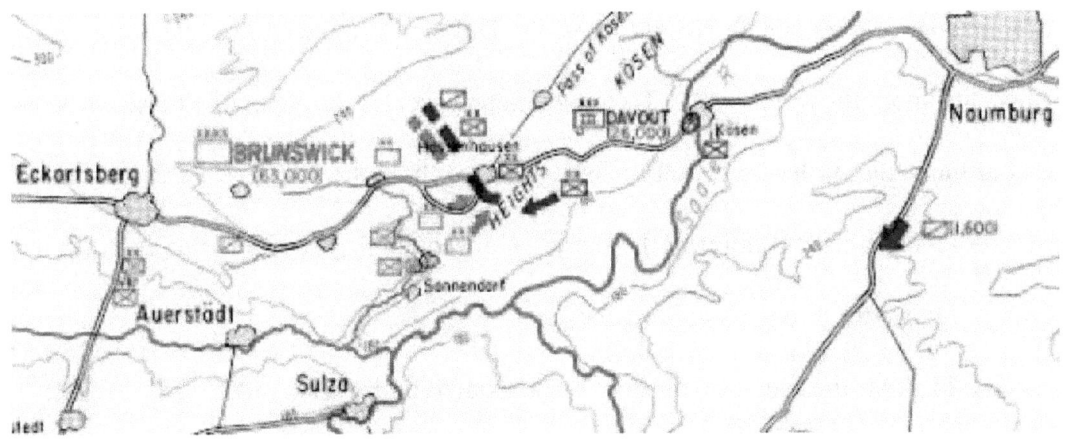

Figure 7. Battle of Auerstadt. Situation 10 A.M., 14 October 1806. Source: United States Military Academy History Department Maps. Available from http://www.dean.usma.edu/history/dhistorymaps/images/Napoleon/nap30.jpg.

Finally, at almost 10 A.M., the Prussians began their advance. The Prussian advance to the north was defeated, but their advance to the south managed to displace the

French. Davout was able to rally his men and retake the village of Hassenhaussen, but in doing so had employed all forces at his disposal.

At this point an interesting event occurred, or more correctly, failed to occur. Bernadotte, with his I Corps of 25,000 men to the south, could have marched to Davaout's relief, but instead chose to remain in position near the town of Dornburg. As a result, Davout's I Corps would completely miss both battles of the campaign. Historians have long argued over the reasons for Bernadotte's reluctance, and many point to the animosity between the two commanders as the cause. Whatever the reason, the fact remains that Bernadotte failed to employ his forces, and consequently came to no ones aid on 14 October 1806.

Just when things looked their worst for Davout, the gods of war smiled and inflicted an unfortunate twist of fate against the Prussians. During the Prussian assault, the commander of the main Prussian forces, the Duke of Brunswick, was shot through the eyes and fell mortally wounded.[40] To make matters worse, the King of Prussia, who was present at the battle, failed to appoint a successor for the duke's loss. The Prussians were now leaderless at a critical point in the battle. Consequently, they failed to exploit their advantage against the French southern flank.

By 11 A.M. the remaining reinforcements arrived for both sides. The French reinforcements of General Morand were committed, in total, to shore up Davout's southern flank. The Prussian reinforcements, however, led by the Prince of Orange, were split in half and committed to the north and south, thus unfortunately negating the possibility of the Prussians weighting one of the flanks. Morand's forces proved decisive and not only repelled the Prussian counterattack, but also succeeded in destroying the

Prussian southern flank.⁴¹ Meanwhile, King Frederick William III refused to release any of the Prussian reserves, for he still believed he was up against Napoleon himself and would need his reserve for a later time. The French continued to press the attack and by half past noon, the Prussians were in full retreat to the west and north. Blucher made a valiant effort to provide a covering screen with his cavalry, but it proved useless. At 4:30 P.M., Davout halted his infantrymen for a much needed rest. Davout continued the pursuit with his three regiments of light cavalry with the intent of harassing the enemy and forcing him to retreat southward towards the French main body. The battle had finally ended for Davout, but the price had been costly. The III Corps sustained over 7,000 casualties, but did manage to kill 10,000 Prussians, capturing another 3,000 along with 115 cannons.⁴² Figure 8 depicts the situation as of 2 P.M., 14 October 1806.

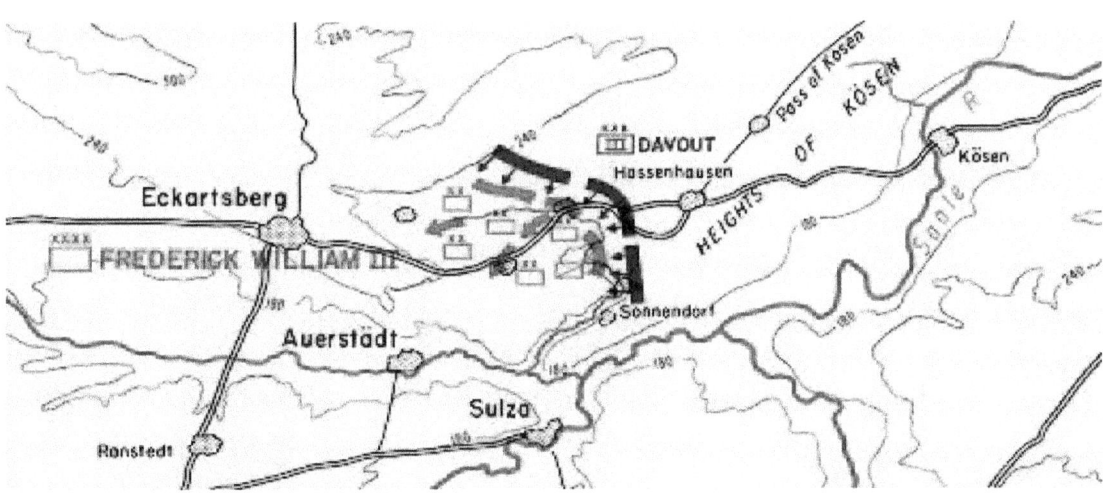

Figure 8. Battle of Auerstadt. Situation 2 P.M., 14 October 1806. Source: United States Military Academy History Department Maps. Available from http://www.dean.usma.edu/history/dhistorymaps/images/Napoleon/nap31.jpg.

When Napoleon was notified of Davout's victory, at 2 A.M. on 15 October, he refused to believe it. Napoleon was still convinced that it was he, and not his subordinate Davout, who had engaged and defeated the Prussian main effort. He informed Davout's messenger that, "Your marshal . . . saw double today."[43] However, by 16 October it was clear that the French forces led by Davout and his III Corps of 26,000 men had in fact engaged the main Prussian force of 64,000 men, while Napoleon, with 96,000 men, had only engaged a smaller force of 55,000 Prussians conducting a shaping operation. General Count de Segur, who was at the time an eyewitness to the event and an aide-de-camp to Napoleon, recalled not only the event but also Napoleon's vanity. De Segur wrote in his memoirs, "The glory was too disproportionate for him [Napoleon] to let go forth to the world, living on fame as he did."[44] Reluctantly, Napoleon eventually accepted the facts and gave credit to Davout for his efforts.[45]

The battle of Auerstadt is another example of the French cavalry's decisive contribution to military operations. Although Davout only had his three regiments of light cavalry, his cavalry succeeded in repelling Prussian attacks throughout the day. Following the battle, his cavalry continued the pursuit, long after the infantry was rendered completely exhausted, and succeeding in forcing the Prussians southward. Together, the efforts of the cavalry at both the battles of Jena and Auerstadt help render Napoleon's victory decisive.

The Chase Begins

On the morning of 15 October, Napoleon ordered his forces to begin a "strategic pursuit" against the retreating Prussian army. As historian David G. Chandler describes, "The scale and ruthlessness of the pursuit that followed the battles of Jena-Auerstadt have often been described, and it provides a classical instance of the way in which a victory can be exploited."[46] Napoleon's plan called for the main French effort to apply continuous pressure against the retreating Prussian forces. This force would assault to delay the enemy's withdrawal in order to enable the French supporting effort to effectively cut off the Prussian line of retreat. The supporting French effort would have the task of seizing the crossing sites across the Elbe River with the purpose of denying the Prussian line of retreat northward toward Berlin or the Oder River.[47]

Murat led the pursuit with his cavalry, followed by Ney's VI Corps and Bernadotte's I Corps. On 16 October, Murat's forces succeeded in capturing the town of Erfurt, which included between 9,000 and 14,000 prisoners (historians differ in their accounts).[48] The same day, Bernadotte's I Corps succeeded in defeating the Prince of Wurtemburg's forces near the town of Halle. The Prussian losses were 5,000 men and eleven cannons, which was almost half of their original number.[49] By 22 October, the French had succeeded in establishing two bridgeheads across the Elbe River and Berlin was firmly within Napoleon's sights. Figure 9 depicts the pursuit to the Oder River from 15 October to 1 November 1806.

Figure 9. Jena Campaign. Pursuit to the Oder. 15 October – 1 November 1806. Source: United States Military Academy History Department Maps. Available from http://www.dean.usma.edu/history/dhistorymaps/images/Napoleon/nap32.jpg.

On 24 October, in recognition of Davout's accomplishments at Auerstadt, his III Corps was given the honor of being the first French unit to march into the Prussian capital of Berlin.[50] Once in Berlin, Napoleon ordered the pursuit to continue. His main concern was the possibility of Russian intervention. He therefore directed Davout to move his III Corps to the east to secure the Oder River. Napoleon continued the pursuit of Prussian forces to the north with the forces of Bernadotte, Lannes, and Murat. With the intent of completely destroying the Prussian army, he directed to Murat, "Let not a man escape!"[51] Hohenlohe was taken prisoner at Prenzlau on 28 October along with 10,000 Prussians and sixty-four cannons.[52] On 29 October, General Milhaud and his 700 cavalrymen accepted the surrender of 4,000 Prussian cavalrymen at Pasewalk; and Lasalle's 5th and 7th Hussars captured the fortress town of Stettin, which contained 5,500 men and 120 cannons.[53] The last Prussian element to surrender was Blucher's who, after being driven from Lubeck by forces from Murat, Bernadotte and Soult, finally surrendered on 7 November on the Danish frontier. With the Prussian army effectively captured or destroyed, the Prussians agreed to an armistice on 26 November. They suffered a loss of over 25,000 killed or wounded and surrendered an additional 140,000 into captivity along with over 2,000 cannons. Only a small number of Prussian forces ever managed to link up with their Russian allies to the east.[54]

This "strategic pursuit" clearly demonstrates the French cavalry's effectiveness at the operational level of war. The major operation, the pursuit, succeeded in accomplishing its strategic objective--the destruction of the Prussian army. The cavalry's mobility, speed, and ability to exploit success proved instrumental in achieving victory. With the cavalry's mobility, the French successfully pursued the Prussians for over 250 miles until their last force surrendered. The cavalry's speed enabled the French to out maneuver the Prussians and capture key terrain, such as towns and river crossing sites, thus denying the Prussians any route of escape. The French cavalry was also able to exploit their success by maintaining a constant pressure upon the Prussian rear guard.

This constant pressure resulted in the deterioration of the Prussian army and forced it into total disarray. The end result of this cavalry focused mission was the total destruction of the Prussian army, rendering Napoleon free to impose his will upon another European nation. Rarely do the conditions exist for an opportunity of this magnitude to occur. However, in this case, the French cavalry was at the right place, at the right time. They were clearly a key element in Napoleon achieving his decisive tactical, operational, and strategic victory over the Prussians.

Analysis

The intent of this analysis is to determine the impact cavalry had on determining the outcome of the Jena-Auerstadt Campaign. For this, each type of cavalry employed, light, medium, and heavy, will be examined within the context of the three levels of war: strategic, operational, and tactical. The analysis will also examine the cavalry's employment in the three types of military operations: decisive, shaping and sustaining.

For the sake of clarity, I will review the definitions of the levels of war and the three types of military operations as I discuss the analysis. The source for these definitions is the United States Army's current Operations Field Manual, FM 3-0, dated June 2001.

The Levels of War

Strategic Level of War. The strategic level is that level at which a nation, often as one of a group of nations, determines national and multinational security objectives and guidance and develops and uses national resources to accomplish them.

Given this definition and applying it to the context of Europe in 1806, there is no particular type of cavalry, or for that manner any specific branch of the military, that necessarily lends itself to a true 'strategic' example. At best, a deterrent force against a

Prussian attack, such as the 18,000 man army of Napoleon's brother, Louis the King of Holland, might apply. Or, in the same context, the strategic threat posed against France by the two 60,000 man Russian armies in the east could apply as well. But in the context of 1806 Europe, no particular branch alone is applicable to this level at this level of war.

It is also interesting to point out that at the strategic level of war, all decisions rested almost entirely with Napoleon himself. He alone was the exclusive campaign planner. He was also the one who directed the allocation of the French national resources, which in this case would be the *Grande Armee* and the supplies required to sustain it.

Operational Level of War. The operational level of war is the level at which campaigns and major operations are conducted and sustained to accomplish strategic objectives within theaters or areas of operations. It links the tactical employment of forces to strategic objectives.

At this level of war it is important to emphasize that campaigns are conducted to sustain strategic objectives. During the Jena Campaign, Napoleon's strategic objective was the destruction of the Prussian army. In this aspect the pursuit phase of the campaign, led by Murat's cavalry, proved essential. The cavalry's mobility, speed, and ability to exploit success proved instrumental in achieving overall victory. A prime example of this is the employment of Lasalle's 5th and 7th Hussars during the pursuit of Prussian forces. Due to their speed and mobility, the Hussars succeeded in capturing the fortress town of Stettin, which contained 5,500 men and 120 cannons. Capturing a superior force of this size and strength, with only light cavalry forces, would not have been feasible had the conditions not been set by Napoleon's exploitation of a defeated enemy. The capture of Stettin demonstrates the excellent ability of the cavalry to exploit success and maintain the initiative established during the execution of a pursuit.

Since the operational level of war also provides the link between the tactical employment of forces and the strategic objectives, the use of cavalry again fits this mission. During the pursuit of the Prussians, the cavalry was not only given the mission of seizing key river crossing sites across the Elbe and Oder Rivers, but also charged with the mission of cutting off the Prussian lines of retreat. Based on these requirements and given the roles and missions of the light and medium (dragoons) cavalry, we see their contributions to the operational level of war.

Tactical Level of War. Tactics is the employment of units in combat. It includes the ordered arrangement and maneuver of units in relation to each other, the terrain, and the enemy to translate potential combat power into victorious battles and engagements.

Since this definition describes the employment of units in combat, it obviously applies to all three types of cavalry. The light and medium cavalry were employed during the reconnaissance and screening missions, as well as being used to maintain contact between advancing columns and securing Napoleon's lines of communication. Likewise, the heavy cavalry was exclusively reserved for the main battle so it could deliver the decisive blow against the enemy at the critical place and time.

The Types of Operations

Decisive Operations. Decisive operations are those that directly accomplish the task assigned by the higher headquarters. Decisive operations conclusively determine the outcome of major operations, battles, and engagements. There is only one decisive operation for any major operation, battle or engagement for any given echelon. The decisive operation may include multiple actions conducted simultaneously throughout the area of operation. Commanders weight the decisive operation by economizing on combat power allocated to shaping operations.

The three critical aspects of a decisive operation are that they conclusively determine the outcome of a battle or engagement, that there is only one decisive operation per battle or engagement, and that the decisive operations are weighted. Given these criteria, the employment of the heavy cavalry during the main battle is an example of a decisive operation. Murat's use of his cuirassiers against Ruchel's forces in the battle of Jena was a decisive operation that determined the outcome of the engagement.

The employment of the light cavalry during the pursuit phase of this campaign also fits the definition of a decisive operation. Since the pursuit of the Prussian forces continued over a period of several days, there were multiple decisive actions conducted over the course of the pursuit. An example of this would be Murat's capture of the town of Erfurt on 16 October. As a result, Murat effectively eliminated any further resistance by the Prussians in that sector.

Shaping Operations. Shaping operations at any echelon create and preserve conditions for the success of the decisive operation. Shaping operations include lethal and nonlethal activities conducted throughout the area of operations. They support the decisive operation by affecting enemy capabilities and forces or by influencing the opposing commander's decisions. Shaping operations use all the elements of combat power to neutralize or reduce enemy capabilities. They may occur before, concurrently with, or after beginning of the decisive operation. They may involve any combination of forces and occur throughout the area of operation.

The role of the cavalry is clearly demonstrated in shaping operations. The cavalry not only created conditions for success, but also engaged in nonlethal combat activities, were able to neutralize the enemy, and continued to shape events after the decisive action had occurred. When the cavalry executed their reconnaissance missions as the lead elements for the advance through the Thuringerwald, they were creating the conditions for success by gaining contact with the enemy. When Lasalle's Hussars made contact with the enemy forces in Leipzig, the cavalry's presence conveyed a nonlethal message to the Prussians that their lines of communication were effectively threatened. Soult's cavalry pressing the attack against the Prussian force at Nerkwitz during the battle of Jena is an example of the cavalry's ability to neutralize the enemy's capability. And since shaping operations can occur either before, concurrent with, or after decisive operations, Davout's cavalry continuing the pursuit of the retreating Prussians following the battle of Auerstadt would serve as an example.

Sustaining Operations. Sustaining operations are operations at any echelon that enable shaping and decisive operations by providing combat service support (CSS), rear area and base security, movement control, terrain management, and infrastructure development.

Cavalry's contribution to sustaining operations came in the form of security missions and movement control. The cavalry was not only employed to sustain Napoleon's lines of communication as he advanced, but they also managed to secure towns along the march routes to ensure the main body's progress was not impeded. With regard to movement control, using the cavalry as guides and to reestablish contact between advancing columns are two examples.

Assessment of the Cavalry

In answering the question of what impact cavalry had on the outcome of the Jena Campaign, their contribution is demonstrated in three areas: finding the enemy, shaping the battlefield, and driving home the decisive victory through an historic pursuit. With regard to the three levels of war, all three types of cavalry made their contributions at the tactical level. However, of all three types of cavalry, it was only the light cavalry in their pursuit role that made a significant contribution at the operational level of war. Thus it was Napoleon's ability to pursue the retreating enemy that allowed him to achieve his strategic objective of destroying the Prussian army.

As for cavalry's contribution to the three types of operations, the heavy cavalry clearly contributed to Napoleon's success in decisive operations. However, only the light cavalry can make the claim of being able to contribute to the French success across all three types of operations. This demonstrates the versatility of this force and sheds possible light for future developmental concerns.

Campaign Insights

The Jena Campaign provides numerous battlefield insights that are applicable to today's commander, namely thorough planning prior to battle, mutually supporting and flexible formations, and the value of the pursuit.

Napoleon's capabilities as a planner, being able to predict approximately a month in advance the location of the expected battle, exemplify what Jomini describes as a reasonable decision reached by "well founded hypothesis." He attributed Napoleon with the ability to, "Make such arrangements that his columns, starting from points widely separated, were concentrated with wonderful precision upon the decisive point of the zone of operations. In this way he insured the successful issue of the campaign."[55] This example reinforces today's efforts in replicating this by conducting the military decision making process as well as executing a thorough Intelligence Preparation of the Battlefield (IPB) prior to an operation.

Napoleon's genius was also demonstrated by his use of the *Battalion Carre* formation as he advanced to make contact with the Prussian army. The *Battalion Carre* not only provided the needed flexibility, but it also afforded the French army the ability to mass against the enemy in only two days time--an extraordinary event for that day in age. This formation proved to be a clever and innovative concept for its time and had a devastating effect against the Prussians.

The last insight is perhaps the one for which the Jena Campaign is best remembered—the pursuit. The Prussian military theorist, Carl von Clausewitz, effectively summarized this when he stated, "Next to victory, the act of pursuit is the most important in war."[56] By Napoleon continuing his pursuit of the Prussian army, he rendered the victory not only decisive, but also complete. By the time the *Grande Armee* had captured the Prussian capital of Berlin, the Prussian army had either been destroyed or was nearly all captured, thus rendering the Prussians helpless and in no position to bargain with the victor. Napoleon's military victory, largely due to his pursuit, had now set the conditions for him to impose his diplomatic will.

We will revisit these insights and analysis again in the final chapter when we draw our final conclusions and apply the trends identified with Napoleon's cavalry to today's transforming United States Army.

---

¹R. Ernest and Trevor N. Dupuy, *The Encyclopedia of Military History from 3500 B.C. to the Present* (New York: Harper and Row, Publishers, 1977), 749.

²Owen Connelly, *Blundering to Glory: Napoleon's Military Campaigns* (Wilmington, Delaware: Scholarly Resources, Inc., 1987), 95.

³Ibid.

⁴Octave Aubrey, *Napoleon: Soldier and Emperor,* trans. Arthur Livingston (Philadelphia: J. B. Lippincott and Company, 1938), 170.

⁵Ibid.

⁶Henry Lachouque, *Napoleon's Battles: A History of His Campaigns*, trans. Roy Monkcom (New York: E. P. Dalton and Company, Inc., 1967), 136.

⁷Connelly, 95.

⁸Lachouque, 136.

⁹Ibid., 139.

¹⁰David G. Chandler, *The Campaigns of Napoleon* (New York: Scribner, 1966), 462.

¹¹Yorck von Wartenburg, *Napoleon As a General*, ed. Walter H. James (London: Kegan Paul, Trench, Trubner and Company, Ltd., 1902), 273.

¹ Lachouque, 136.

²Chandler, *The Campaigns of Napoleon,* 464.

³Aubrey, 171.

⁴Wartenburg, 280.

⁵Ibid., 273.

⁶Lachouque, 139.

⁷Chandler, *The Campaigns of Napoleon*, 468.

⁸Wartenburg, 285.

⁹Chandler, *The Campaigns of Napoleon*, 469.

[10] Wartenburg, 287.

[11] Ibid.

[12] Ibid., 288.

[13] Lachouque, 140.

[14] Wartenburg, 289.

[15] Lachouque, 141.

[16] Louis-Nicolas Davout, "Correspondence of Marshal Davout, Prince of Eckmuhl," *The War Times Journal* 12 October 1806, 177 [Online] Available from http://www.wtj.com/archives/davout/, 10 February, 2002.

[17] Lachouque, *Napoleon's Battles*, 141.

[18] Ibid.

[19] Ibid., 142.

[20] Ibid.

[21] Wartenburg, 291.

[22] Lachouque, 144.

[23] Wartenburg, 292.

[24] Chandler, *The Campaigns of Napoleon*, 478.

[25] Wartenburg, 293.

[26] Lachouque, 144.

[27] Chandler, *The Campaigns of Napoleon*, 477.

[28] Ibid., 480.

[29] Ibid., 481.

[30] Ibid., 483.

[31] Ibid.

[32] Ibid.

[33] Ibid., 484.

³⁴Ibid.

³⁵Ibid., 485.

³⁶Ibid., 488.

³⁷Ibid.

³⁸Ibid., 490.

³⁹Ibid.

⁴⁰Ibid., 492.

⁴¹Ibid., 494.

⁴²Ibid., 495.

⁴³Connelly, 101.

⁴⁴General Count de Segur, *An Aide-de-Camp of Napoleon*, trans. H.A. Patchett-Martin (London: Hutchinson and Company, 1895), 291.

⁴⁵Connelly, 102.

⁴⁶Chandler, *The Campaigns of Napoleon*, 497.

⁴⁷Ibid., 498.

⁴⁸Ibid.

⁴⁹Ibid.

⁵⁰Connelly, 103.

⁵¹Lachouque, 150.

⁵²Chandler, *The Campaigns of Napoleon*, 501.

⁵³Ibid.

⁵⁴Ibid., 502.

⁵⁵Antoine Henri Jomini, *The Art of War*, ed. Brig. Gen. J. D. Hittle (Mechanicsburg, Pennsylvania: Stackpole Books, 1987), 535.

⁵⁶Carl von Clausewitz, *Principles of War*, trans. and ed. Hans W. Gatzke (Mechanicsburg, Pennsylvania: Stackpole Books, 1987), 337.

CHAPTER 4

THE SAXONY CAMPAIGN

The purpose of analyzing Napoleon's Saxony Campaign of 1813 is to gain further insights on the impact cavalry forces, or if any, the lack of cavalry forces had on determining the outcome of a campaign. To accomplish this, the historical background of the campaign will be reviewed along with a chronological description the events that occurred. The study will also conduct an analysis of cavalry's contribution in determining the outcome of the campaign. Cavalry forces will be considered using the construct of the three levels of war, as well as the three types of operations. This study will also attempt to gain possible insights from the Saxony campaign that might be applied towards current mounted operations. Insights to tactical procedures and force structure considerations will be sought out as well.

Strategic Setting

Napoleon's Saxony Campaign of 1813 was a matter of survival for the French Empire. Militarily, the Grand Armee had been devastated during its retreat from Moscow in the failed Russian Campaign of 1812. Although the exact numbers of the French losses are not known, historian David Gates records the following: 570,000 soldiers, 1,050 cannons, and 200,000 cavalry troops and horses.[1] Of these losses, the reduced numbers of quality cavalry troopers proved detrimental in the Saxony Campaign.

Politically, France's allies were becoming restless and wavering with regard to their loyalty to the Emperor. Things were especially troublesome in Prussia, Germany, and the surrounding area. Prussia ended its alliance with France in December 1812 by signing the Convention of Tauroggen,[2] the Confederation of the Rhine was showing signs of fracture, the loyalty of Saxony was in question, and whether or not Austria would remain in a neutral state became suspect. All of this was compounded by the fact France remained in a state of war with Russia; England continued its naval blockade against France and its Peninsula campaign against French forces in Spain. Marshal Bernadotte, who was once loyal to Napoleon, was now the Crown Prince of Sweden and was eager to join forces with the Allies in hopes to increase Sweden's territorial gains. Thus, Napoleon now faced the challenge of rebuilding his army while also maintaining the delicate status quo in Europe, where France still controlled the most power and attempted to enforce its weakening economic policy of the Continental System. For Napoleon, France's course of action was clear: he would once again seize the initiative and launch a military campaign against the enemies of the empire to teach them a lesson. As French historian Henry Lachouque describes, "The only way out of a situation which was daily becoming more unbearable was through victory, which would silence the enemies and bring round the waverers."[3] The critical waverer, in this case, was Austria with her army of over 127,000 men.[4]

The *Grande Armee*

Napoleon's first task was to regenerate lost combat power. To accomplish this he relied on existing French troops to form his core army and from which he built his new *Grande Armee*. These men came from various garrisons not affected by the Russian Campaign, such as the harbor and municipal garrisons. Because Napoleon lost fourteen *cuirassier* regiments in his Russian invasion, he was now forced to pulling dragoon regiments from Spain and refit them as heavy cavalry units. Together, these garrisons and veterans of Spain accounted for approximately 100,000 men.[5] To further augment this, Napoleon relied on soldiers from his para-military National Guard as well as vast numbers of raw recruits. By the spring of 1813, Napoleon's forces once again numbered around 745,000.[6] Although, very young and inexperienced, the regenerated *Grand Armee* did not lack in zeal and enthusiasm. Only the soldiers from his foreign contingents of Germany, Italy, Holland, Switzerland, and Croatia had to be persuaded from defecting to the Allies through the use of force.[7]

Although Napoleon was able to compensate for his loss of men, compensating for his loss of horses and cavalry troopers proved more challenging. First of all, the question of suitable horses became an issue. As previously discussed, the most suitable horses for the cavalry came from either Normandy or Prussia and central Germany. Requisitioning additional horses from Normandy would not be a problem, provided there was a sufficient quantity. However, trying to requisition horses from Prussia and central Germany proved increasingly difficult since their national alliances and political preferences were leaning toward the Allies.[8] The matter of quickly regenerating combat power was further frustrated by the limited time available to train the new cavalry troopers. Cavalry troopers took considerably longer to train, when compared to their infantry brethren, due to the complexity of their missions and precision required to adequately perform their battle drills. Consequently, the problems of an insufficient quantity of cavalry horses and inadequately trained cavalry troopers would have their impact on Napoleon's Saxony Campaign of 1813. Not only would he suffer from inadequate reconnaissance reporting and cavalry screening, but Napoleon would also lack sufficient cavalry forces to pursue a defeated foe off the battlefield and render the victory decisive. Table 3 depicts the French order of battle for the Saxony campaign as of October 1813.

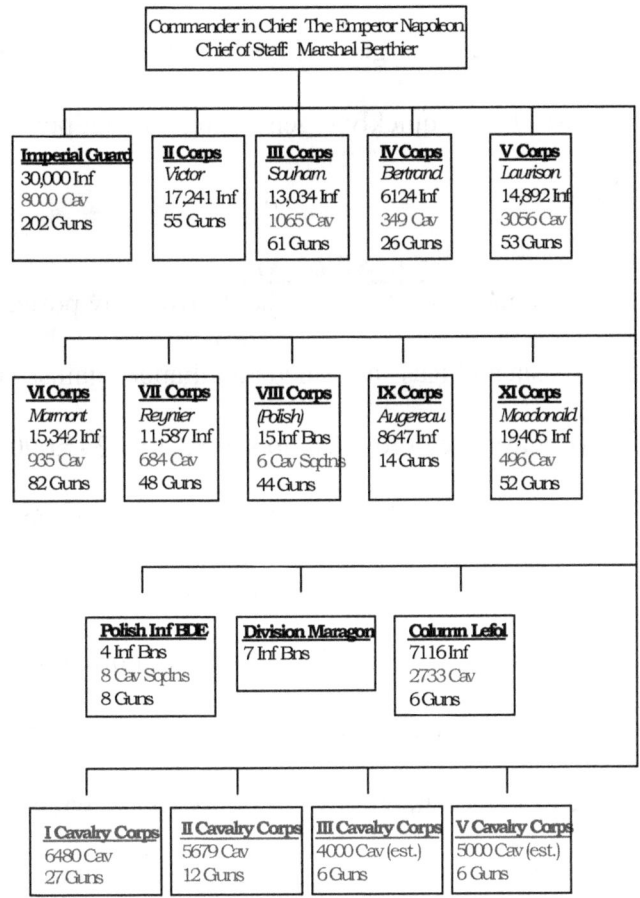

Table 3. French Order of Battle for the Saxony Campaign October, 1813. Source: Peter Hofschroer, *Leipzig 1813, The Battle of the Nations* (Oxford, Osprey Publishing Ltd., 1993), 28-30.

The new *Grand Armee* had other deficiencies as well. The French artillery was plagued by the same shortage of horses the cavalry experienced, and the quality of

French officers began to differ greatly. The senior officers, such as colonels, who were carefully selected usually proved to be excellent. However, the junior officers, although eager to serve, were not of the same high caliber. They often lacked physical stamina and were simply not seasoned enough due to their rapid rate of promotion from the masses.[9] Although Napoleon's new army was larger, the quality of this new *Grand Armee* was not the same as that of the past. For in quickly regenerating combat power, Napoleon had traded away quality for quantity.

The Allied Army

In comparison to the *Grand Armee*, the Allied army posed a credible threat. On 28 February 1813, Prussia and Russia formed an alliance against France by signing the Treaty of Kalisch. Together they provided approximately 345,000 troops. Austria, who was initially neutral, officially joined the Allies on 27 June by signing the Treaty of Reichenbach along with Russia and Prussia. When Austria later declared war against France on 12 August, she increased the allied numbers by approximately 127,000 troops. Sweden landed forces on the continent in Pomerania on 18 May, and officially joined the coalition on 21 July. She brought with her over 23,000 troops, which when combined with the other forces, including Anglo-Germans and Mecklenburgers, gave the Allied coalition over 500,000 troops.[10]

Although the Allies succeeded in fielding sufficient numbers to challenge Napoleon, they were plagued by an age-old dilemma of coalition warfare—unity of effort. Napoleon, the common enemy, appeared to be the only thing holding the coalition together. Russia was determined to rid Europe of Napoleon. Prussia demanded revenge for their 1806 losses at Jena as well as the disintegration of the Confederation of the

Rhine. Austria, according to historian David Gates, "Wanted to counterbalance France and Russia and restore the Austro-Prussian hegemony in Germany."[11] And Sweden was interested in freeing Norway from Danish control. These varying agendas proved a hindrance for the first half of 1813 and Napoleon would not fail to capitalize on the Allies' inability to act quickly. However, the second half of 1813 would be a different matter for the Allies when their combined strategy proved effective against Napoleon's *Grand Armee*.

The Battle of Lutzen

In the spring of 1813, Napoleon decided to take the initiative against the coalition that was forming to oppose France. His objectives were to push the Russians back behind their own frontiers, regain control of Berlin and the Prussian government, and recover the French soldiers who were isolated in their garrisons at Danzig (30,000 soldiers) and Stettin (9,000 soldiers).[12] To achieve this, Napoleon devised a plan similar to the one he had used seven years earlier at Jena. On 12 April, Napoleon described his ideas in a note to General Bertrand: "It is my intention to withhold my right and allow the enemy to penetrate to Bayreuth, with a movement opposite to that which I made in the Jena campaign, so that when the enemy has advanced on Bayreuth, I can get to Dresden and cut him off from Prussia."[13] With this in mind, Napoleon advanced east towards Leipzig along two axes, one from northern Germany, led by Prince Eugene, the other from central Germany, led by himself. Together the French had approximately 200,000 soldiers and 370 cannons at their disposal, but with only 15,000 cavalry troopers this force still lacked the quantity of cavalry that Napoleon desired. Consequently, Napoleon was not only advancing somewhat blindly into enemy territory, but he was also deprived of a sufficient cavalry screening force to prevent the enemy from gaining the exact location of the French forces. As Napoleon described in a letter to the King of Wurttemberg, "I would find myself in a position to finish matters very quickly if I only had 15,000 more cavalry."[14] As historian David Gates describes, "The feebleness of Napoleon's mounted forces in the German campaign was to have both tactical and strategic repercussions, for success on the battlefield could not be consolidated and exploited by the sort of pursuit *a outrance* he had so frequently staged in the past."[15]

However, things were far from perfect in the Allied camp. Czar Alexander continued to meddle in the affairs of his army staff and undermined the authority of his commanders. Russian General Kutusov died on 28 April and the Czar was forced to divide the command between himself and General Wittgenstein. This, combined with the other complications of waging war by coalition, left the Allied headquarters in a frustrated state of affairs.[16] With regard to the training of the Allied forces, the Prussians had the advantage of a well-trained core of professional infantrymen and the Russians enjoyed the benefit of a large number of veteran soldiers. Although the rapidly increased numbers of new recruits diluted the quality of both armies to a degree, the Allies and the French were about on par with regard to the quality of their soldiers.

As the French forces advanced towards Leipzig, during the end of April, one noteworthy incident occurred. Napoleon suffered the loss of one of his distinguished cavalry leaders, Marshal Bessieres. During one of the skirmishes prior to the battle of Lutzen, Bessieres' cavalry had been sent into action to reinforce General Kellermann's cavalry forces as they countered the Russian cavalry. However, during the fight, Bessieres was killed when a cannon ball ricocheted off of a nearby wall.[17] His death came as a severe blow to Napoleon and he was without a key subordinate for his campaign.

As Napoleon maneuvered towards Leipzig, with approximately 120,000 men, he was aware that his lack of cavalry reconnaissance left him vulnerable. Without sufficient reconnaissance, Napoleon was without accurate intelligence that would enable him to determine the enemy disposition and strength. If he could not determine the enemy's disposition and strength, Napoleon remained at a disadvantage when deciding whether or not to give battle against the enemy. To mitigate his lack of reconnaissance forces, Napoleon directed Marshal Ney to occupy four villages southeast of Lutzen: Kaja, Rahna, Gross and Klein Groschen, as a precautionary measure incase the Allies attacked from the town of Zwenka in the east. In the event Ney did encounter the enemy, Napoleon planned for Ney's forces to fix the enemy while the main French force enveloped the southern Allied flank.[18] Unfortunately for Napoleon, Ney did not heed his direction of, "Send(ing) out two strong reconnaissance forces, one toward Zwenken (in the east) and the other toward Pegau (in the south)."[19] Instead, Ney only committed two of his five divisions to occupy the villages to the southeast and left the remaining three divisions near the town of Lutzen. Consequently, the French remained weak and vulnerably exposed with only a small force in the town of Kaja.

Meanwhile, Allied cavalry reconnaissance discovered this French weakness. Their plan of attack called for Allied forces to conduct a movement to contact, from east to west, in order to establish contact with the small French forces at Kaja and then to proceed westwardly in order to interdict and cut off the road running from Weissenfels to Lutzen, along which French forces were advancing.[20]

On 2 May, the two sides made contact and the battle of Lutzen began as approximately 73,000 Allied forces, led by Russian General Wittgenstein, attacked into the French flank against Ney's III Corps of 40,000 men.[21] The initial French forces were arrayed west to east covering a front approximately a mile long. The Allies were established along a one and a half mile front, attacking the French from south to north. General Blucher, with the Allied main effort, was in the east and the Allied cavalry was positioned in the west. Figure 10 depicts the Battle of Luetzen.

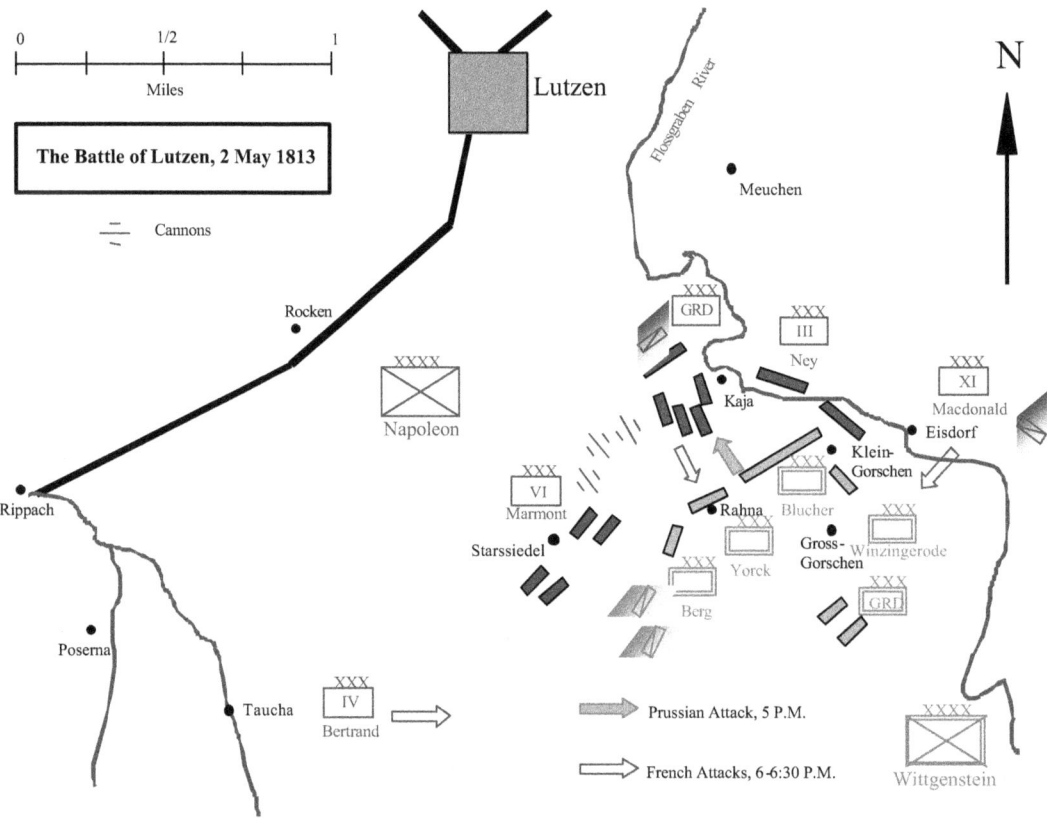

Figure 10. Battle of Lutzen, 2 May 1813. Source: David G. Chandler, *The Campaigns of Napoleon* (New York: Scribner, 1966), 885.

Although the Allied headquarters had issued long and complex orders to their forces, the Prussians managed to take up positions on the unsuspecting French forces. Blucher's Prussian forces began the attack with cavalry and artillery, believing they would quickly sweep the French from the battlefield. However, the French resistance was fierce and a determined fight ensued. Allied artillery succeeded in pushing the French center back, but Ney countered this by ordering his three uncommitted divisions into the fight.

Napoleon, who was in the historic town of Lutzen, where Gustavus Adolphus had fallen in battle in 1632, heard the attack and rode southeast to the sound of the guns. Upon his arrival, Napoleon personally took charge and directed the initial French forces, of only 45,000 men, into battle. As historian David Chandler records, "The effect of his presence was almost magical."[22] Napoleon decided to fix the Allied forces in the center while Marshal Macdonald's XI Corps would attack the Allied eastern flank and Marshal Marmont's VI Corps and Marshal Bertrand's IV Corps would attack the Allied western flank. Napoleon positioned his reserve force, the Imperial Guard, in the center sector near the village of Kaja.[23]

The fight was long and bloody. Blucher was wounded and General Yorck had to assume his command. The Russian reserves were slow to get into position and by 4 P.M. the Allies had committed all of his reserve forces into the battle. By 5:30 P.M., the French succeeded in positioning their forces on the enemy's flanks and at 6 P.M., Napoleon launched his counterattack. Napoleon had massed his artillery in the center sector and four battalions of Young Guards spearheaded the counterattack. The attack proved successful and recaptured the villages of Rahna, Klein and Gross Groschen. Once

Wittgenstein realized what had occurred, he ordered an organized withdrawal from the battlefield. The fighting ended when darkness fell, but the lack of French cavalry deprived Napoleon of his pursuit *a l'outrance*. With inadequately trained and insufficient numbers of cavalry forces, Napoleon was unable to attack the enemy when they were at their most vulnerable point, in a retreat. Marmont's exhausted infantry continued to pressure the enemy, but they were effectively countered by Prussian cavalry attacks. As Chandler writes, "The result was undoubtedly a victory for Napoleon, but the inadequacy of the pursuit robbed him of complete success."[24]

The French victory had cost both sides valuable men. Both the Allies and the French sustained approximately 18,000 casualties each and the Prussian General Scharnhorst, who served as Wittgenstein's chief of staff, eventually died from an infection contracted from wounds received during the battle.

The battle of Lutzen had demonstrated the Allies' improved capabilities. The Russians had fought determinedly, and the Prussians had proven themselves a changed army since Napoleon had routed them from the battlefield of Jena seven years earlier. It was following the battle of Lutzen when Napoleon exclaimed, "These animals have learned something."[25] To add to his frustrations, the question of Austria's neutrality still remained. Therefore, Napoleon would have to seek another victory in an attempt to sway Austria to the French side.[26]

For the French, however, one thing stood out—the lack of cavalry. This lack of sufficient cavalry had haunted Napoleon twice during the battle. The first time was prior to the battle, when the cavalry failed to correctly gain and maintain contact with enemy forces. Consequently, Ney's forces were caught by surprise and the large Allied force threatened Napoleon's southern flank. The second time was following the battle, when the lack of sufficient cavalry prevented Napoleon from executing an effective pursuit. This inability to pursue resulted in the French achieving only a limited, rather than a decisive, victory. Unfortunately for Napoleon, this would not be the last time his lack of cavalry would plague him

The Battle of Bautzen

The Allies continued their withdrawal to the east, but Napoleon's lack of cavalry left him almost completely in the dark with regard to the enemy's exact location. Consequently, to maintain pressure on the Allies, Napoleon split his force into two elements and sent one north toward Berlin and the other, under his command, towards Bautzen. Ney was in command of the northern French force with 79,500 men and 4,800 cavalry troopers. Napoleon selected Macdonald as his deputy and was in charge of 110,000 men and 12,000 cavalry troopers, most of which was predominantly light cavalry. In order to reestablish contact with the enemy, Napoleon ordered a reconnaissance in force, led by Macdonald, with the VI, XI, and IV Corps.[27]

Meanwhile, the curse of the coalition had again taken effect as political considerations started dictating military actions. The Czar, concerned with protecting the Russian lines of communications wanted his forces withdrawn to the east towards Breslau. The Prussians, concerned with the protection of Berlin, wanted their forces withdrawn to the north.[28] With dissention in the Allied camp, Napoleon continued his drive eastward and succeeded in capturing Dresden on 8 May. After the fall of Dresden, the Allies finally decided to counter Napoleon's advance and massed their forces near the town of Bautzen on 16 May. Figure 11 depicts the disposition of French and Allied forces arrayed for the Battle of Bautzen on 20 May 1813.

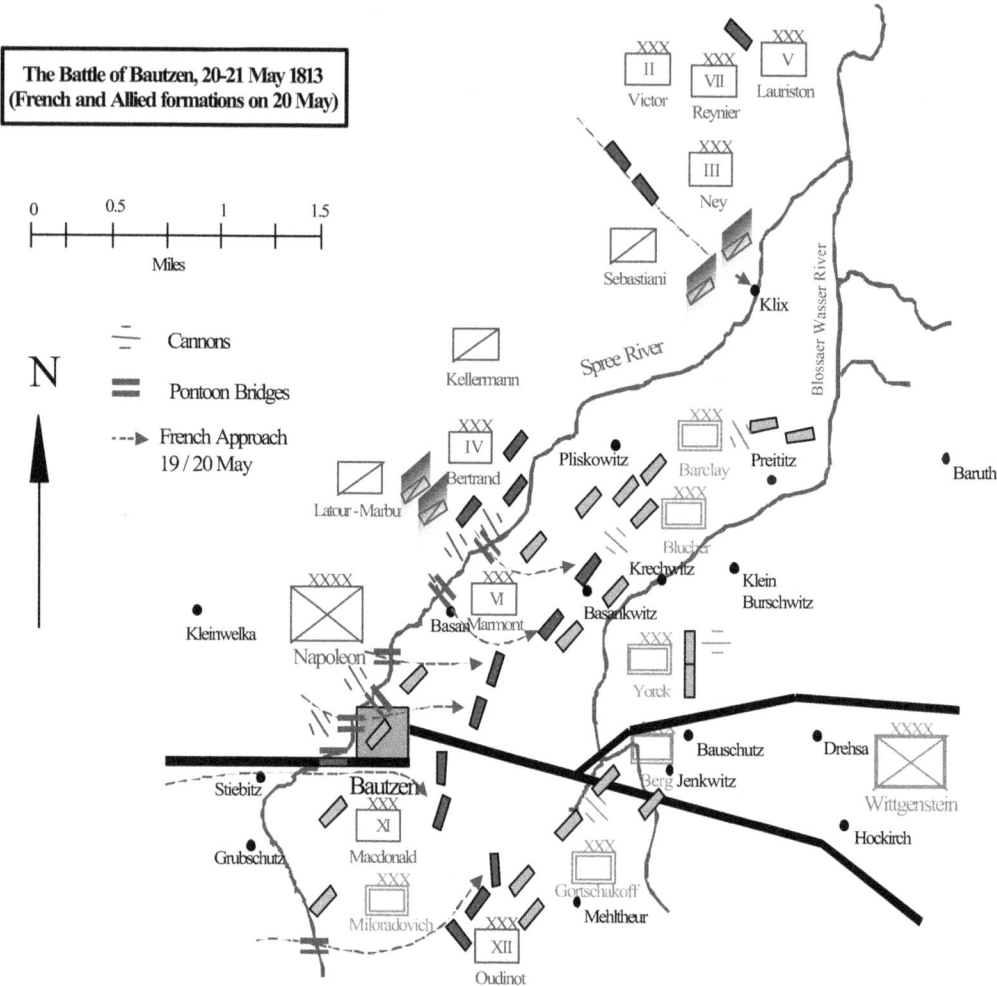

Table 11. Battle of Bautzen. Source: David G. Chandler, *The Campaigns of Napoleon* (New York: Scribner, 1966), 892.

For this battle, French forces were arrayed along a front, approximately five miles long, running southwest to northeast along the west bank of the Spree River. Consequently, the Allies took up positions arrayed along the eastern bank of the river.

Since the French forces were split into two elements, Napoleon decided to inflict a hammer and anvil type maneuver against the Allies. Three of his Corps, Marmont's VI, Macdonald's XI, and Bertrand's IV, would fix the enemy in the center sector. Marshal Oudinot and his XII Corps would conduct a feint to the south in an attempt to draw Allied forces towards him; and Ney's northern force, consisting of his III Corps, followed by Reynier's VII Corps, would attempt to envelop the enemy form north to south. Napoleon would leave his Imperial Guard in reserve in the center sector. Critical to this entire plan was Ney's ability to attack from the north and cut off the Allied route of retreat to the east. This would enable Napoleon to envelop the Allied army and defeat it in detail.[29]

Historians have criticized Napoleon's decision to use Marshal Ney to command the northern French maneuvering force since the very capable Marshal Davout, of Auerstadt fame, was available. However, Napoleon had decided instead to utilize Davout in capturing the northern German town of Hamburg at the beginning of the Saxony Campaign. Had Napoleon selected Davout instead of Ney, things might have turned out differently.[30]

Napoleon envisioned the battle of Bautzen lasting two days. The first day would be a battle of attrition in the center sector while Ney's forces positioned themselves to the north. The second day would be the battle of envelopment. The Allied plan intended to hold the French in check and exhaust them, while the Allies prepared to counter attack from the north.[31]

Napoleon spent the entire day of 19 May conducting reconnaissance in preparation for the impending battle. Through his reconnaissance efforts he was able to determine the enemy's approximate strength and disposition. The Allies also took advantage of this tactical pause to send out a strengthened reconnaissance force of their own to assess the French. Bertrand responded by sending out an Italian division to repel the Allied reconnaissance force, but failed miserably. The Allies would have succeeded in obtaining further information about French forces had it not been for Kellermann's cavalry division, and Ney's advance guard, that finally succeeded in repelling the Allied force.[32]

The battle of Bautzen began on 20 May with the Allies successfully massing 96,000 men and 622 cannons.[33] Napoleon's fixing force advanced eastwards and threatened the Allied southern flank. Ignoring reconnaissance reports that an attack might come from the north, the Czar committed his reserve troops to shore up his threatened flank to the south. By the end of the first day, Napoleon had achieved his objectives by fixing the Allied force in the center and also firmly establishing French bridgeheads east of the Spree River. The success in gaining the bridgeheads was due to the valiant efforts of the French engineers who constructed several pontoon bridges, often times while under fire from the enemy. These bridges allowed the French forces to rapidly penetrate the river line and seize more advantageous terrain to the east. The next day would be Ney's opportunity to deliver the swift and decisive blow from the north that would encircle the Allied army and provide Napoleon his decisive victory that had evaded him at the battle of Lutzen.

The morning of 21 May began as planned for the French as Ney's advanced guard crossed the river by 5 A.M. Crucial to this plan, Napoleon had ordered Ney to have his forces seize the town of Preititz by 11 A.M. in order to enable the French to envelop the enemy and cut off the Allied lines of retreat. However, rather than proceeding rapidly, Ney hesitated and allowed himself to become decisively engaged with Blucher's forces, instead of bypassing them and cutting off the Allied line of retreat to the east. As Ney's chief of staff, Baron Antoine Henri Jomini later recalled, "If Ney had executed better what he was advised to do, the victory would have been a great one."[34] Fortunately for Napoleon, the French had at least succeeded in fixing the Allied forces to the south and the Czar, incorrectly, dismissed Ney's approach from the north as a diversionary attack. Slowly, Napoleon continued his advance eastwardly from the center, but the attack stalled since Ney had not been able to secure the town of Preititz. Consequently, the Allied escape routes remained open. Fortunately for the Allies, the Czar realized what was happening and ordered an organized withdrawal eastward towards Silesia.[35] A torrential rainstorm ended the battle around 10 P.M., and the Allies continued their withdrawal. Morning found the Allies completely gone and the French too exhausted to continue the pursuit. Once again the lack of French cavalry compounded the situation. Both sides suffered approximately 20,000 casualties, and Napoleon had been forced to commit his valuable reserve to achieve his limited victory.[36]

The battle of Bautzen had once again reinforced the French need for sufficient cavalry. Had cavalry been available, Napoleon would have had the opportunity to turn his limited victory into a decisive one by pursuing and destroying the retreating Allied forces. This lack of cavalry would be one of the reasons Napoleon consented to an armistice with the Allies in June 1813.

The battle at Bautzen affected the Allies as well. This defeat was the second in only three weeks and had shaken their resolve. The Allies began to argue over what to do next. The Prussians desired to remain and fight the French on German soil for obvious political reasons. However, the Russians wanted to withdraw further to the east, into Poland, so they could safely regroup and refit. Eventually, the Czar succeeded in brokering an agreement and the Allied army agreed to move into Silesia, near the town of Schweidnitz.[37]

The Armistice

On 2 June 1813, the Allies and France agreed to an armistice at Pleiswitz, which eventually extended into August. In a note to his minister of war, Napoleon gave his reasons for consent. He stated, "This armistice will interrupt the course of my victories. Two considerations have made up my mind: my shortage of cavalry, which prevents me from striking great blows, and the hostile attitude of Austria."[38] Twice in the same month, Napoleon had been deprived of decisive victories due to a lack of cavalry, and now there were approximately 150,000 Austrian troops massing near Prague that he would have to contend with. Napoleon hoped to take advantage of the growing rift between Russia and Prussia and ultimately return to the battlefield with regenerated French forces. His intent was to contain the Russians while he concentrated his forces against defeating the Prussians. Even if Austria intervened, he still believed it possible to eliminate the Prussian threat before Russia or Austria could come to her aid.[39]

It is interesting to note that Napoleon specifically cited his lack of cavalry as one of the reasons for his agreeing to the armistice. This clearly demonstrates not only the cavalry's contribution to Napoleon's efforts, but also his reliance upon their capabilities. Napoleon had developed a method of warfare that was dependent on gaining accurate knowledge of the enemy strength, location, and disposition through effective reconnaissance. In the battle of Lutzen, insufficient cavalry forces allowed him to be taken by surprise. In the battle of Bautzen, Napoleon had to dedicate an entire day to reconnaissance before he committed his men into battle. Both instances indicate that an accurate picture of the enemy's location and disposition was crucial to Napoleon developing his course of action. Without sufficient numbers of well-trained cavalrymen, this did not occur. The other interesting cavalry capability that was lacking was the ability to follow up the battlefield victory with an effective pursuit. Napoleon's method of warfare relied on his cavalry to finish the destruction of the enemy force if it retreated from the battlefield. Without the cavalry's ability to effectively pursue the foe, Napoleon's victories were not decisive as in previous campaigns.

Both sides looked forward to the opportunity to regenerate lost combat power, and in essence this period became something of an arms race. Throughout Europe, nations committed to one side or the other. With regard to France, she succeeded in swaying Bavaria and Denmark into her camp as well as gaining the support of the King of Naples, Marshal Ney.

As for the Allies, the armistice proved to ultimately be in their favor since they could compensate for their losses at a faster rate than France. On 15 June, Prussia and Russia agreed to a subsidized arrangement with Great Britain whereby Russia and Prussia would continue their struggle against France until she was crushed. Great Britain agreed to pay Russia thirty-three million pounds a month, and Prussia seventeen million pounds a month for their continued efforts.[40] As if this wasn't enough, Sweden officially joined the coalition on 21 July and Austria declared war on France, citing France's inability to agree to the peace terms as reason for cause. On 16 August the hostilities resumed.[41] For the first time since 1795, France stood alone against all the powers of continental Europe.[42]

The Road to Leipzig

Where Napoleon had been on the offensive prior to the signing of the armistice in June, the situation now forced him to assume a posture of strategic defense. He now faced four formidable armies: The Army of Bohemia, led by Austrian Prince von Scwarzenberg, with 240,000 soldiers; the Army of the North, led by the Crown Prince of Sweden Bernadotte, with 120,000 soldiers; the Army of Silesia, led by Prussian Prince Blucher, with 95,000 soldiers; and the Army of Poland, led by Russian Marshal Bennigsen, with 60,000 soldiers. All totaled, the Allied numbers eventually reached over 600,000 men.[43]

As for France, Napoleon was able to gather three armies, led by Marshals Ney, Oudinot, and himself, but could only field around 250,000 soldiers. To make matters worse, the French army continued to reduce in size for each additional battle it fought.[44] At least Napoleon had two things in his favor: central positioning and centralized command. He therefore decided to seize the tactical initiative and attack his most dangerous enemy first, Blucher, while keeping the others in check.[45]

However, the Allies had developed a strategy that if successful, would prove Napoleon's undoing. Their new plan, named the Trachenberg Plan, called for the Allies to seek a series of limited objectives rather than immediately seeking the total destruction of the French army. The plan had six key points: "One, any fortresses occupied by the enemy were not to be besieged but merely observed. Two, the main effort was to be directed against the enemy's flanks and lines of operations. Three, to cut the enemy's communications, forcing him to detach troops to clear them or move his main forces against them. Four, to accept battle only against part of the enemy's forces and only if that part were outnumbered, but to avoid battle against his combined forces, especially if these were directed against the Allies' weak points. Five, in the event of the enemy moving in force against one of the Allied armies, this was to retire while the others advanced with vigor. And six, the point of union of the Allied armies was to be the enemy's headquarters."[46]

For the next three months, August through October, Napoleon attacked the Allies in vain. The Allies held to their Trachenberg Plan and denied Napoleon the opportunity to decisively engage an Allied army. They succeeded in avoiding battle with Napoleon himself, and instead chose to battle and destroy the forces under the command of his subordinates.[47]

As Napoleon launched his attack against into Silesia against Blucher, on 21 August, the Austrian led Army of Bohemia attacked the French forces remaining near Dresden under command of St. Cyr. When Blucher retreated from Napoleon in the east, Napoleon turned back to meet the Army of Bohemia near Dresden. No sooner had Napoleon departed and given part of his command to Macdonald, than Blucher reappeared in the east and successfully defeated Macdonald's forces at the Katzbach River on 26 August.[48]

On 27 August, Napoleon's force of 120,000 men and 250 cannons met with and attacked the Army of Bohemia consisting of 170,000 men and 400 cannons. The French center held while they attempted a double envelopment around both flanks of the enemy. On the night of 27-28 August the Allies held to their plan and withdrew under the cover of darkness in order to avoid decisive engagement. This relatively hollow victory for Napoleon at Dresden would be his last on German soil. For once again, the enemy had retreated and Napoleon lacked the cavalry forces necessary to follow in pursuit.[49]

Meanwhile, both to the north and south of Dresden, Napoleon's subordinates were being defeated. To the north, Oudinot's French force had been defeated at the battle of Grossbeeren on 23 August. To the south, French General Vandamme, had been defeated and captured after a valiant fight during the battle of Kulm. On 6 September, Marshall Ney took command of the northern French forces, from Oudinot, and went on the offensive against Bernadotte. Ultimately, Ney was soundly defeated at the battle of Dennewitz. Meanwhile Napoleon attempted to engage both Blucher and Schwarzenber on two separate occasions, but both Allied armies avoided battle.[50]

The Allies continued to apply direct pressure against Napoleon from the north, east, and south, thus threatening his lines of communication with France to his west. On 24 September, Napoleon ordered his forces west of the Elbe River in order to reduce his front and protect his extended lines.[51]

The Battle of the Nations

By mid-October, all of the Napoleon's forces were concentrated in the vicinity of Leipzig. The Allies had succeeded in their Trachenberg Plan and now had Napoleon surrounded on three sides, to the north, east, and south, with his only means of escape remaining to the west through Leipzig. The battle of Leipzig, also referred to as the Battle of the Nations, was essentially a French retrograde operation in the face of an overwhelming enemy that lasted from 16 to 19 October. Napoleon, with 177,000 men and 700 cannons, faced four armies totally 410,000 men and 1,500 cannons. It was only a matter of time until Napoleon was forced out of Germany.[52]

As Napoleon attempted to consolidate his forces in the vicinity of Leipzig, he ordered Murat to use the French cavalry to delay the Army of Bohemia that was advancing on Leipzig from the southeast. On 14 October, Murat fought a series of cavalry engagements against the Allied forces of Wittgenstein that resulted in the largest cavalry battle of 1813, the battle of Liebertwolkwitz.

This battle consisted of a series of cavalry charges that were answered by counter-charges. The counter-charges were then followed by a pursuit that was led by the reserve forces. The commitment of the reserve forces would then compel the opponent to commit his reserve forces, which usually resulted in the termination of the engagement. It is interesting to note that by October 1813 the Allies had figured out an effective countermeasure to the French cavalry charge. Because the French preferred to attack in column formation, the Allies countered this by fixing the lead elements of the French column and then launching a counterattack against the flank of the French column before it could effectively deploy to defend itself.[53]

The battle of Liebertwolkwitz ended inconclusively. The French succeeded in delaying the advance of the Army of Bohemia, but lost valuable cavalry forces in the process. Although the exact number of losses is not certain, it is estimated that both sides lost over 2000 men and over 600 horses.[54] This battle did, however, alert Napoleon to the fact that an Allied attack on Leipzig was now imminent.[55]

On 16 October, Napoleon attacked Schwarzenberg's Austrian forces to the south. Concurrently, Blucher's Prussian forces engaged the French to the north. Both sides fought aggressively, but no territory was gained on either side.

No significant fighting occurred on 17 October, but reinforcements did arrive for both sides. Reynier's Corps of 18,000 men reinforced the French, while Bennigsen's Russians, with 70,000 men, and Bernadotte's Swedes, with 85,000 men, reinforced the Allies. Given this situation, Napoleon concluded that a retreat to the west was the only course of action remaining.[56] Figure 12 depicts the array of French and Allied forces for the Battle of Leipzig as of 19 October 1813.

On 18 and 19 October, the fighting raged to the north, east, and west of Leipzig as the Allies assaulted the French forces. The Allied assault against the city continued on 19 October as the French proceeded with their orderly retreat. However, at 1 P.M. a nervous French corporal prematurely detonated the bridge leading out of the city. Consequently, four French corps were trapped in the town. The corps fought bravely, but to no avail. By the end of the day, the French had suffered approximately 68,000 casualties along with an additional 30,000 Frenchmen being taken prisoner. The Allies also suffered heavily with 54,000 casualties.[57]

Thus, the battle of Leipzig marked the end of Napoleon's rule in Germany. The Bavarians, who had originally sided with Napoleon, switched their allegiance to the Allies, as did the Saxons, who actually defected from France during the battle of Leipzig. Napoleon later recalled this event in his memoirs when he wrote, "Thereupon the whole Saxon Army, with sixty guns, which was occupying one of the most important positions in the army, went over to the enemy and turned their guns on the French line. An act of treachery of this unheard-of kind was bound to bring about the ruin of the French Army, and give the honor of the day to the Allies."[58] As Napoleon continued his retreat to the west, additional minor German rulers switched their allegiances to the Allies as well. By 7 November, Napoleon had departed his *Grand Armee* to return to Paris. On 8 November, the Allies offered Napoleon terms for peace: Napoleon could remain in power provided he ruled over France limited by its natural boundaries--the Rhine River and the Alps. Napoleon refused this offer and the Allies invaded France on 1 December 1813.[59]

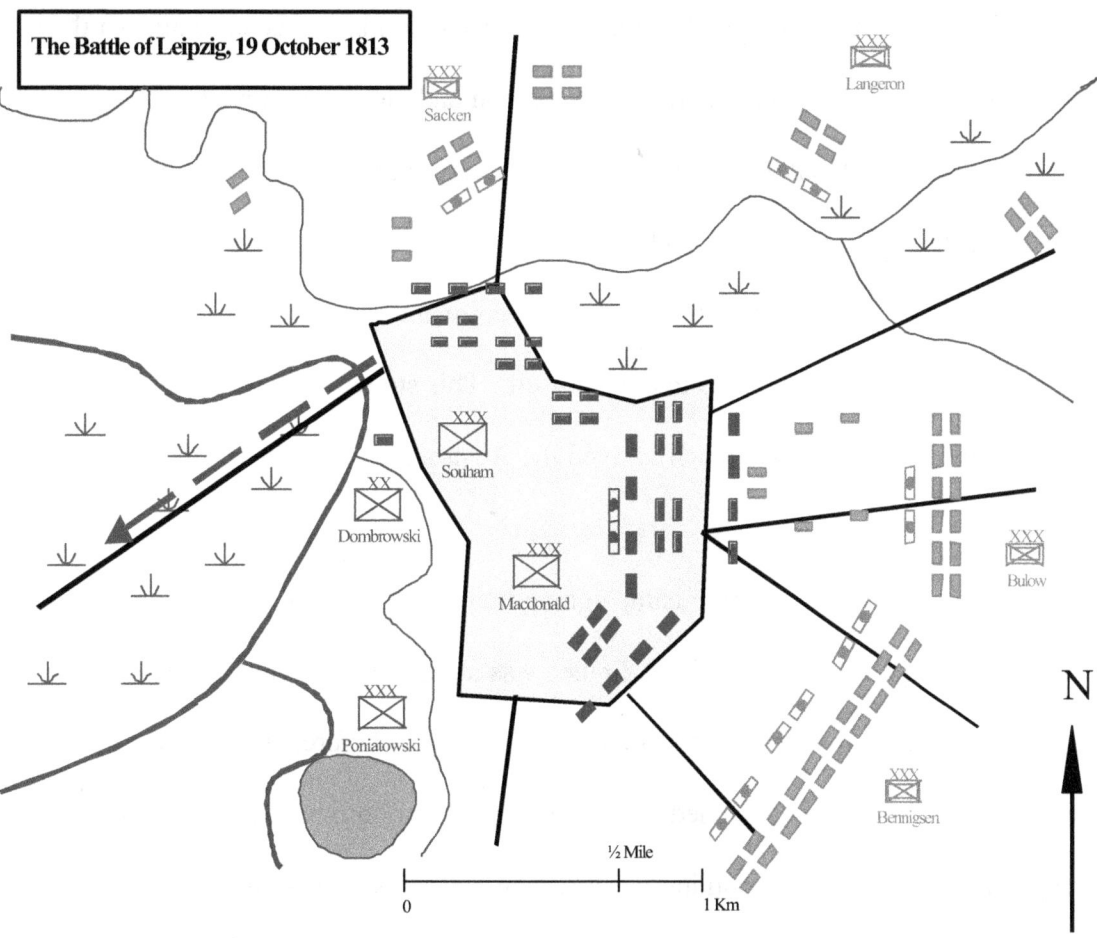

Figure 12. Battle of Leipzig. Source: Peter Hofschroer, *Leipzig 1813, The Battle of the Nations* (Oxford: Osprey Publishing Ltd., 1993), 87.

Conclusion

The Saxony Campaign of 1813 proved to be the watershed event for both the Allies and Napoleon. Operationally, the Allies had finally succeeded in forming a viable coalition that was not only able to outnumber the quantity of soldiers France could produce, but it was also finally able to match them in quality. By 1813, the Allies had substantial numbers of battle proven soldiers, especially in the armies of Prussia and Russia, who also succeeded in having a superior cavalry force, both in numbers and quality. The Allies also demonstrated their ability to learn from their mistakes and even adopted several of the tactics and techniques that Napoleon had once used against them. The Allies were finally a formidable foe. Napoleon had taught them well.

Strategically, the Allies forced their coalition to work. The greatest evidence of this was the development of the Trachenberg Plan. They pitted their strengths against Napoleon's weaknesses, namely his subordinate commanders, and avoided decisive engagements when up against the master himself. This strategy, combined with successful diplomatic efforts, which swayed the Austrians to the Allied side, proved effective.

As for France, the years of campaigning had finally taken their toll, especially the Russian Campaign of 1812. Although France was able to rapidly generate an army in a short period of time, it was not the same *Grande Armee* as before. Tactically, France no longer had the numbers of seasoned infantrymen or junior commanders as she once had. Strategically, France was consuming her forces in battle faster than they could be produced. However, of all the items an army could want for, it would be cavalry that would have its most telling effect.

During the first half of 1813, the lack of cavalry affected Napoleon in three ways: it denied him an adequate means of reconnaissance, it denied him an adequate screening force, and it denied him the ability to pursue a driven foe from the battlefield. Consequently, Napoleon did not have a good read on the enemy prior to battle, could not adequately deny the enemy the knowledge of the location of his forces, and could not achieve decisive victories for lack of his ability to pursue the enemy. All of these factors came to bear and in June, Napoleon agreed to the armistice citing lack of cavalry as one of his reasons. After June, although Napoleon recovered to a certain degree, he had reached the point of diminishing returns against an Allied coalition that grew stronger each day. Napoleon had no other option but to either risk the rest of his empire or sue for peace.

Analysis

The intent of the analysis for this chapter is to once again assess the impact cavalry had on determining the outcome of the campaign. I will follow the same construct as in the previous chapter and analyze each type of cavalry within the context of the three levels of war and the three types of military operations.

The Levels of War

Strategic Level of War. During the Saxony Campaign, Napoleon's strategic objectives were to push the Russians back behind their own frontier borders, regain control of Berlin and the Prussian government, and liberate the isolated French garrisons of Danzig and Stettin. Napoleon failed to accomplish all three of these objectives. When attempting to assess the cavalry's impact at the strategic level, as previously discussed, there is no particular type of cavalry, or for that manner any specific branch of the military, which necessarily lends itself to a true 'strategic' example when reviewing the Napoleonic Wars.

Operational Level of War. At the operational level of war, Napoleon was twice denied a decisive victory during the Saxony Campaign of 1813. These denials occurred following the battles of Lutzen and Bautzen and were directly caused by the lack of cavalry, which prevented him from conducting a pursuit of the retreating enemy. There is perhaps no stronger endorsement for the need of cavalry to help secure victory than when Napoleon specifically mentioned his lack of cavalry as one of the reasons for consenting to the armistice in June 1813. As historian David Chandler has described, "The general shortage of cavalry horses made effective exploitation impossible."[60] Without sufficient cavalry Napoleon remained at an operational disadvantage.

Tactical Level of War. With regard to the tactical level of war, Napoleon's lack of cavalry was felt in three distinct areas: reconnaissance, security, and the pursuit. Without adequate cavalry forces to conduct reconnaissance, Napoleon was in the dark with regard to the enemy's disposition. This was demonstrated during the battle of Lutzen when the Allied army caught Ney's III Corps by surprise on 2 May. The battle of Lutzen also provides an example of Napoleon's inability to screen his own force movements and prevent the enemy from gaining an understanding of the French forces disposition. Wittgenstein was fully aware of the limited French forces southeast of Lutzen and also that the road from Lutzen to Weissenfels was a main logistics route for the French. With sufficient cavalry, Napoleon could have established a more effective screening force and denied the enemy this valuable information. And as for the pursuit, as mentioned earlier, Napoleon was unable to drive home the victory following a battle due to insufficient cavalry forces.

The Types of Operations

Decisive Operations. Napoleon's inability to pursue the retreating enemy limited his decisive operations to the battlefield itself. In the case of the Saxony Campaign, the single element that consistently proved decisive for Napoleon was his Imperial Guard that he held in reserve. In both battles of Lutzen and Bautzen, the commitment of the Guard proved decisive and caused the enemy line to recoil and eventually retreat. Unfortunately, Napoleon's lack of cavalry denied him the ability to effectively pursue his retreating foe.

Shaping Operations. With sufficient forces, the role of the cavalry is usually demonstrated in shaping operations, such as reconnaissance and screening missions. However, without it is more difficult for the commander to set the conditions for success on the battlefield. As was demonstrated at the battle of Bautzen, Napoleon directed his corps to conduct a "reconnaissance in force" since his cavalry reconnaissance forces were lacking. Likewise, Napoleon was forced to rely on pickets to provide early warning against an enemy advance, since he lacked sufficient cavalry to execute an adequate screen line. When the pickets were not sufficiently posted, as was Ney's case at the battle of Lutzen, this deficiency was amplified even further.

Sustaining Operations. Typically cavalry's contribution to sustaining operations are in the form of security missions and movement control. Securing Napoleon's lines of communications proved challenging during the Saxony Campaign. There were several cases of enemy cavalry raids, particularly by Cossacks, against the French lines of communication and supply bases. Once again, if Napoleon had been able to generate the required amount of cavalry forces, the enemy would not have enjoyed their freedom to maneuver behind the French lines. As for movement control, although specific incidents were not cited, one can assume that insufficient cavalry would have had its affect on this mission as well.

Assessment of the Cavalry

The Saxony Campaign demonstrates the fact that Napoleon's cavalry had been a key element to achieving a decisive victory. French cavalry played an important role before, during, and after Napoleon's battles. With sufficient and well-trained cavalry, Napoleon's victories were decisive; without it they were hollow. Before the battle, the light and medium cavalry served critical roles in shaping the battlefield. They were charged with providing accurate reconnaissance reports on the enemy as well as seizing key terrain. They also screened the movement of the main body, thus preventing the enemy from knowing Napoleon's true disposition. All of these actions helped shape the battlefield. During battle, Napoleon's heavy cavalry was typically held in reserve to be committed at the critical place and critical time to deliver the decisive blow against the enemy. Thus, the cavalry's commitment to the main battle proved decisive. Following the main battle, light cavalry was used to pursue elements of the foe and complete total destruction of the enemy's force, which was habitually Napoleon's objective. Therefore, with sufficient and well-trained cavalry, Napoleon's victories were decisive, as in the Jena-Auerstadt Campaign of 1806; without it they were hollow, or at best Pyrrhic, as in the Saxony campaign of 1813. Clausewitz summed up the value of the pursuit when he stated, "Only the pursuit of the beaten enemy gives the fruits of victory."[61]

Campaign Insights

History has demonstrated that every campaign has the ability to teach enduring lessons that are applicable to all generations. Napoleon's Saxony Campaign of 1813 is no different. Three insights from this campaign that are applicable for today's commander are the value of seasoned troops, the importance of understanding the commander's intent, and the value of cavalry on the battlefield.

After Napoleon's catastrophic invasion of Russia in 1812, the *Grande Armee* was decimated. Although he tried to regenerate the French army in less than six months, he traded valuable quality for quantity in order to be able to go on the offensive in the spring of 1813. Consequently his new troops, though they fought valiantly, lacked the stamina and skill that his pre-Russian *Grande Armee* demonstrated. They were slower to maneuver and lacked critical skills, such as the ability to conduct a thorough reconnaissance. Although seemingly minor, these issues would take their cumulative effect on the success of the campaign.

Another valuable insight is the ability for a subordinate commander to understand and apply his higher commander's intent during battle. For with this level of understanding, subordinate commanders can conduct decentralized operations more successfully. Without this understanding, the operation stands a more difficult chance of succeeding. This point was clearly demonstrated at the battle of Bautzen when Ney failed in his mission of cutting off the Allied line of retreat to the east. Consequently, the Allied army escaped and Napoleon's chances of a decisive victory were further frustrated by his inability to launch an effective pursuit. Had Ney completely understood Napoleon's purpose and intent for his enveloping attack, it is likely that Ney would not have allowed his forces to become decisively engaged before seizing the point of blocking the Allied escape route.

Finally, the Saxony Campaign demonstrates a prime example of the necessity of cavalry on the battlefield. With sufficient cavalry, the commander can conduct thorough reconnaissance missions and help shape the battlefield to his favor. He can also turn a limited tactical victory into a decisive victory through an effective pursuit. In this regard, the Saxony Campaign stands as the antithesis of the success Napoleon had achieved at Jena seven years earlier.

---

[1]David Gates, *The Napoleonic Wars, 1803-1815* (London: Arnold, 1997), 221.

[2]Peter Hofschroer, *Leipzig 1813: The Battle of Nations*, ed. David G. Chandler (Oxford: Osprey Publishing, Ltd. 1993), 91.

[3]Henry Lachouque, *Napoleon's Battles: A History of His Campaigns*, trans. Roy Monkcom (New York: E.P. Dalton and Company, Inc. 1967), 330.

[4]Hofschroer, 37.

[5]Ibid., 9.

[6]Ibid., 11.

[7]Lachouque, 329.

[8]David G. Chandler, *The Campaigns of Napoleon* (New York: Scribner, 1966), 868.

[9]Lachouque, 328.

[10]Hofschroer, 27.

[11]Gates, 225.

[12]Lachouque, 332.

[13]Ibid.

[14]Gates, 235.

[15] Ibid., 233.

[16] Ibid., 235.

[17] Lachouque, 335.

[18] Chandler, *The Campaigns of Napoleon*, 881.

[19] Ibid.

[20] Ibid., 882.

[21] Owen Connelly, *Blundering to Glory: Napoleon's Military Campaigns* (Wilmington, Delaware: Scholarly Resources, Inc., 1987), 184.

[22] Chandler, *The Campaigns of Napoleon*, 884.

[23] Ibid.

[24] Ibid., 887.

[25] Ibid.

[26] Lachouque, 339.

[27] Chandler, *The Campaigns of Napoleon*, 889.

[28] Gates, 239.

[29] Chandler, *The Campaigns of Napoleon*, 889.

[30] Gates, 239.

[31] Chandler, *The Campaigns of Napoleon*, 893.

[32] Ibid., 891.

[33] Gates, 239.

[34] Antoine Henri Jomini, *The Art of War*, ed. Brig. Gen. J. D. Hittle (Mechanicsburg, Pennsylvania: Stackpole Books, 1987), 507.

[35] Gates, 241.

[36] Connelly, 187.

[37] Gates, 242.

[38] Chandler, *The Campaigns of Napoleon*, 898.

[39] Gates, 243.

[40] Lachouque, 349.

[41] Hofschroer, 91.

[42] Gates, 245.

[43] Connelly, 189.

[44] Ibid.

[45] Hofschroer, 37.

[46] Ibid., 40.

[47] Connelly, 190.

[48] Ibid., 191.

[49] Ibid.

[50] Ibid.

[51] Ibid., 192.

[52] Ibid., 193.

[53] Hofschroer, 69.

[54] Ibid., 70.

[55] Chandler, *The Campaigns of Napoleon,* 922.

[56] Connelly, 193.

[57] Ibid.

[58] Napoleon Bonaparte, *Memoirs of Napoleon I*, comp. F. M. Kircheisen, trans. Frederick Collins (New York: Duffield and Company, 1929), 203.

[59] Connelly, 194.

[60] Chandler, *The Campaigns of Napoleon,* 896.

[61] Carl von Clausewitz, *Principles of War*, trans. and ed. Hans W. Gatzke (Mechanicsburg, Pennsylvania: Stackpole Books, 1987), 351.

CHAPTER 5

CONCLUSION

The idea for this study began with an interest in Napoleon's cavalry and the U.S. Army's current struggle with the transformation process in which it is trying to remain strategically relevant for the twenty-first century. Consequently, the study developed into a detailed review of Napoleon's cavalry in order to gain insights for employing current mounted forces, as well as possibly identifying doctrinal or capability trends that might be relevant to the IBCT. The results have been enlightening. This chapter will restate the critical research questions presented in the first chapter and will answer each in a concise narrative.

<u>What contribution did Napoleon's cavalry make towards his military operations?</u>

Cavalry clearly played a decisive role in the execution of Napoleon's military operations before, during, and after the battle. In fact, as demonstrated in the Saxony Campaign, the lack of sufficient cavalry directly affected Napoleon's ability to make sound and correct decisions on the battlefield and exploit tactical opportunities when they appeared.

Before the operation, the light cavalry was critical in shaping the battlefield through effective reconnaissance operations. The medium cavalry, or dragoons, also played an important role in shaping the battlefield by seizing key terrain in order to gain a positional advantage over the enemy. These two actions, reconnaissance and seizing key terrain, provided Napoleon with a clear situational understanding and allowed him the ability to position his forces in order to compel the enemy to have to fight in two directions at once, a critical technique to his battlefield success.

During the military operation, Napoleon's heavy cavalry clearly proved to be a decisive element on the battlefield. Through the execution of a cavalry charge, for either a deliberate attack or penetration of enemy lines, the heavy cavalry often succeeded in breaking through the enemy defense. This success enabled the French to either seize or maintain the initiative. With their lines broken and the initiative lost, the enemy was often compelled to retreat, thereby ending the battle.

After the battle, Napoleon would commit the light cavalry to a pursuit in order to maintain constant pressure against the retreating foe. Through pursuit operations, Napoleon could render his tactical victory decisive through the destruction of the enemy center of gravity--his field army. Depending on how complete the destruction of the enemy force was, it could have a decisive operational or strategic impact.

How did Napoleon organize and employ his cavalry forces with regard to the metric of decisive, shaping, and sustaining operations?

Napoleon employed his cavalry forces in accordance with their capabilities. The organization into light, medium, and heavy cavalry allowed him to apply this arm across the spectrum of operations.

For decisive operations, the heavy cavalry was specifically held in reserve in order to deliver the crucial blow against the enemy at the decisive point and time. Following an operation, the light cavalry was often committed to execute a pursuit mission with the objective of destroying the enemy field army. This may also be considered a decisive operation.

For shaping operations, Napoleon typically relied on the light cavalry to conduct reconnaissance missions and the dragoons to seize key terrain. Both light and medium cavalry also conducted flank security during the battle, which would fall in the category of a shaping operation.

For sustaining operations, the light cavalry was employed to maintain security for the lines of communication. This duty included escorting supply convoys as well as patrolling and maintaining the supply routes clear of enemy forces.

<u>Did Napoleon employ his cavalry differently depending on the level of war he was focused (strategic, operational, tactical)?</u>

Quite simply, yes. Strategically, in the context of Europe in the early nineteenth century, there is no particular type of cavalry, or for that manner any specific branch of the military, which necessarily lends itself to a true 'strategic' example that is applicable to this level at this level of war.

Operationally, Napoleon fought campaigns or major operations to achieve strategic objectives. A typical strategic objective for Napoleon was the destruction of the enemy field army, a center of gravity in nineteenth century warfare. This strategic objective was usually obtained through the tactical victory on the battlefield. If the battlefield victory was decisive, and succeeded in destroying the enemy force, then it had a strategic impact. The pursuit of the Prussian army following the Jena Campaign is an example of actions conducted at the operational level of war to obtain a strategic objective. Once Prussia's field army was destroyed, Prussia had no means to resist Napoleon from imposing his will.

Tactically, Napoleon initially focused his cavalry efforts on conducting shaping operations and then transitioned them to perform decisive operations. The light cavalry shaped the battlefield through reconnaissance. The medium cavalry helped Napoleon gain positional advantage by seizing key terrain. And the heavy cavalry was directed to deliver the decisive blow, or coup de grace, at the critical time and place on the battlefield. Following the battle, the light cavalry was committed to execute a pursuit against any remaining enemy forces or seize decisive terrain in support of follow on operations.

<u>Are there any operational or tactical techniques that may be applied to today's battlefield, or that should be considered for today's force structure?</u>

This question can be effectively answered if it is broken down into two aspects: capabilities and techniques. With regard to military capabilities, being able to shape the battlefield through effective reconnaissance, delivering the decisive blow to the enemy at the critical place and time, and pursuing a retreating enemy are the capabilities that Napoleon's cavalry performed during his military operations. The need to perform these same capabilities transcends time and remains applicable on today's battlefield.

The IBCT's reconnaissance, surveillance and target acquisition (RSTA) squadron is a major breakthrough in Army force development with regard to future reconnaissance capabilities. The RSTA squadron still serves as the brigade's eyes and ears, and according to the IBCT Organizational and Operational Concept, dated 30 June 2000, the RSTA squadron also, "provides a great deal of the information and intelligence required by the commander and staff to do proper planning and, direct operations, and visualize the battlefield."[1] This is achieved through the synchronized efforts of the squadron's recce troops and surveillance troop, equipped with unmanned aerial vehicles (UAV), ground sensor radars, and a Nuclear Biological Chemical (NBC) Reconnaissance platoon, working to develop the required situational awareness and knowledge for the commander in the brigade's area of operation. Once fielded, the RSTA squadron has the potential to prevent campaign fiascos, like the one Napoleon experienced in Saxony in 1813, from happening again. This breakthrough in reconnaissance capabilities should be capitalized on. The U.S. Army would be wise to consider fielding a RSTA squadron to each of its divisions and also insisting the squadron's capabilities be matched, if not surpassed, for any future combat reconnaissance system designs.

The ability to deliver the decisive blow at the critical place and time still remains one of the unique capabilities performed by armor units on today's battlefield. The M1 series tank still has the firepower, maneuverability, survivability, and shock effect that is unmatched by any other combat force. However, there are drawbacks to modern heavy forces. Typically, their deployability requires a longer time than other forces and they also require considerable logistic support for sustainment. These are only two among several reasons why the Army is pursuing an Interim Force until the Objective Force can be developed and fielded to compensate for these shortfalls. Regardless of what the new Objective Force looks like in terms of equipment, the ability to deliver the decisive blow at the critical place and time will remain a crucial capability requirement.

With regard to employment techniques, Napoleon's cavalry performed several methods that are applicable to today's forces, four of which are outstanding: gaining positional advantage, forcing the enemy to fight in two directions at once, threatening the enemy's LOC, and executing a pursuit. By understanding the location of enemy units and seizing key terrain and denying its use from enemy forces, Napoleon gained positional advantage over his opponents. This positional advantage often forced the enemy to have to fight in two directions at once. Once enemy forces were decisively engaged in two different directions, Napoleon committed the heavy cavalry to conduct a penetration at the apex of the enemy line. Once the line was broken, Napoleon then commenced to defeat the two weaker enemy elements in detail. These techniques remain relevant today.

Another technique was seizing key terrain, such as river crossing sites, bridges, and towns that allowed Napoleon to threaten his enemy's LOCs and cut off the enemy line of retreat. This is relevant for today's battlefield as well. By threatening the enemy's LOCs, a commander can often compel the defending unit to dislocate from his prepared positions in order to pull his forces in closer to secure his interior lines. Once the enemy does this, he surrenders the initiative to the other side.

Perhaps the most intriguing technique of all was the ability to execute a pursuit, which often resulted in the total destruction of the enemy field army. This is intriguing because pursuit operations are less common today than they were in the nineteenth century. This may be explained by the lethality of today's battlefield that often achieves the destruction of the enemy force, thus alleviating the requirement for a pursuit. Another factor may also be the increased emphasis on achieving a desired political endstate, rather than simply a military one. In this situation, political objectives outweigh military objectives and therefore do not afford the opportunity to execute a pursuit mission. An example of this is the 1991 Gulf War where the coalition forces were directed to not continue the pursuit and destruction of the retreating Iraqi forces. The political objective of liberating the nation of Kuwait had been achieved, and the continued destruction of Iraqi forces may have led to further complications with regard to political and coalition concerns. Consequently, the pursuit and destruction of the Iraqi forces was called off.

What are the insights from the two campaigns that may be applied to mounted operations today?

There are two insights from these campaigns that are enduring: the value of understanding the commander's intent and the contribution cavalry (or on today's battlefield mounted forces) makes to the outcome of an operation.

With regard to understanding the commander's intent, the actions of Davout at the battle of Auerstadt is a prime example. Although outnumbered by an enemy force more than twice his size, Davout understood his task and purpose and effectively blocked the Prussian line of retreat. He seized the initiative early in the battle and successfully held his ground for over six hours. Had Davout not understood Napoleon's intent, it is very probable that the Prussians would have succeeded in their retreat towards Berlin.

In contrast to Davout's action at the battle of Auerstadt is Ney's action at the battle of Bautzen. At Bautzen, Ney hesitated in conducting his enveloping movement and also allowed himself to become decisively engaged along the enemy's front line trace, rather than penetrating deep into the enemy's rear to cut off his line of retreat. From this performance, one can speculate whether or not Ney had a firm grasp on the commander's intent. If he had, the battle most likely would have turned out differently, resulting in a more thorough defeat of the Allied army.

The other insight gained from this study is the enduring value of cavalry on the battlefield. With sufficient and well-trained cavalry forces, Napoleon was clearly able to shape the battlefield through effective reconnaissance, deliver decisive blows with his heavy cavalry during the battle, and could follow up the victory with a pursuit if the situation required. No other branch of the service could contribute to his operation in quite the same manner. Napoleon's cavalry was truly a key element to decisive victory.

---

[1]Department of the Army, *The Interim Brigade Combat Team Organizational and Operational Concept*, Version 4.0, Final Draft (Fort Leavenworth: USACGSC, 30 June 2000), 36.

BIBLIOGRAPHY

Primary Sources

Bonaparte, Napoleon. *Memoirs of Napoleon I.* Compiled by F. M. Kircheisen and transcribed by FrederickCollins. New York: Duffield and Company, 1929.

Clausewitz, Carl von. *Principles of War.* Translated and edited by Hans W. Gatzke. Mechanicsburg, Pennsylvania: Stackpole Books, 1987.

Davout, Louis-Nicolas. "Correspondence of Marshal Davout, Prince of Eckmuhl." *The War Times Journal.* 12 October 1806, 177. Online. Available from http://www.wtj.com/archives/davout/. Accessed on 10 February, 2002.

de Segur, General Count. *An Aide-de-Camp of Napoleon.* Translated by H .A. Patchett-Martin. London: Hutchinson and Company, 1895.

Jomini, Antoine Henri. *The Art of War.* Edited by Brigadier General J .D. Hittle. Mechanicsburg,Pennsylvania: Stackpole Books, 1987.

U.S., Department of the Army. *The Interim Brigade Combat Team Organizational and Operational Concept.* Version 4.0, Final Draft. Fort Leavenworth, KS: USACGSC, 30 June 2000.

Secondary Sources

Books

Aubrey, Octave. *Napoleon: Soldier and Emperor.* Translated by Arthur Livingston. Philadelphia: J. B. Lippincott and Company, 1938.

Atteridge, A. Hilliard. *Joachim Murat.* London: Methuen and Company, Ltd., 1911.

Breene, R. G., Jr. *The Jena Campaign, Interim Report #1.* Reno, Nevada: Physical Studies, Inc., 1967.

Britt III, Albert Sidney. *The Wars of Napoleon.* Wayne, New Jersey: Avery Publishing Group, Inc., 1985.

Chandler, David G. *The Campaigns of Napoleon.* New York: Scribner, 1966.

_____. *Dictionary of the Napoleonic Wars.* New York: Simon and Schuster, 1993.

_____. *Jena, 1806: Napoleon Destroys Prussia.* Oxford: Osprey Publishing Ltd., 1993.

Connelly, Owen. *Blundering to Glory: Napoleon's Military Campaigns.* Wilmington, Delaware: Scholarly Resources, Inc., 1987.

Dupuy, R. Ernest, and Trevor N. *The Encyclopedia of Military History from 3500 B.C. to the Present.* New York: Harper and Row Publishers, 1977.

Elting, John Robert. *Swords Around a Throne.* New York: The Free Press, 1988.

Gates, David. *The Napoleonic Wars, 1803-1815.* London: Arnold, 1997.

Grbasic, Z., and V. Vuksic. *The History of Cavalry.* New York: Facts on File, 1989.

Griess, Thomas E. *Atlas for the Wars of Napoleon.* Wayne, New Jersey: Avery Publishing Group, Inc., 1986.

Haythornwaite, Phillip J. *The Napoleonic Source Book.* New York: Facts on File, 1990.

Hofschroer, Peter. *Leipzig 1813: The Battle of Nations.* Edited by David G. Chandler. Oxford: Osprey Publishing, Ltd., 1993.

Lachouque, Henry. *Napoleon's Battles: A History of His Campaigns.* Translated by Roy Monkcom. New York: E. P. Dalton and Company, Inc., 1967.

Luvaas, Jay. *Napoleon on the Art of War.* New York: The Free Press, 1999.

Maycock, Captain F. W. O. *Napoleon's European Campaigns, 1796-1815.* London: Gale and Polden, Ltd., 1910.

Muir, Rory. *Tactics and the Experience of Battle in the Age of Napoleon.* New Haven, Connecticut: Yale University Press, 1998.

Oman, Sir Charles. *Studies in the Napoleonic Wars.* London: Methuen and Company, Ltd., 1914.

Petre, F. Loraine. *Napoleon's Conquest in Prussia 1806.* London: Greenhill Books, 1993.

Pope, Stephen. *Dictionary of the Napoleonic Wars.* New York, NY: Facts on File, 1999.

Rothenberg, Gunther E. *The Art of Warfare in the Age of Napoleon.* Bloomington, Indiana: Indiana University Press, 1978.

Strachen, Hew. *European Armies and the Conduct of War.* London: George Allen and Unwin, 1983.

United States Military Academy. *Summaries of Selected Military Campaigns--The Napoleonic Wars.* West Point, New York: Department of Military Art and Engineering, 1968.

U.S., Department of the Army. FM 3-0, *Operations.* Washington, DC: Government Printing Office, 2001.

Wartenburg, Yorck von. *Napoleon As a General.* Edited by Walter H. James. London: Kegan Paul, Trench, Trubner and Company, Ltd., 1902.

<u>Articles in Journals and Magazines</u>

Picard, Ernest. "Maxims and Opinions of Napoleon on the Use of Cavalry." *Journal of the U.S. Cavalry Association* 24 (July 1913 to May 1914): 994-1003.

Niderost, Eric. "The Invincible Image Shattered." *Military History Magazine,* Feb 1989, 18-25.

Simmons, Bowen. "Thunder at Leutzen." *Strategy and Tactics Magazin,* January-February 1985, 16-43.

INITIAL DISTRIBUTION LIST

1. Combined Arms Research Library
   U.S. Army Command and General Staff College
   250 Gibbon Ave.
   Fort Leavenworth, KS 66027-2314

2. Defense Technical Information Center/OCA
   8725 John J. Kingman Rd., Suite 944
   Fort Belvoir, VA 22060-6218

3. MAJ Steven J. Rauch
   Combat Studies Institute
   USACGSC
   1 Reynolds Ave.
   Fort Leavenworth, KS 66027-1352

4. Dr. Thomas Huber
   Combat Studies Institute
   USACGSC
   1 Reynolds Ave.
   Fort Leavenworth, KS 66027-1352

5. LTC John Cantlon
   Department of Tactics
   USACGSC
   1 Reynolds Ave.
   Fort Leavenworth, KS 66027-1352

6. Mr. Mike Browne
   Combined Arms Research Library
   U.S. Army Command and General Staff College
   250 Gibbon Ave.
   Fort Leavenworth, KS 66027-2314

CERTIFICATION FOR MMAS DISTRIBUTION STATEMENT

1. <u>Certification Date</u>: 31 May 2002

2. <u>Thesis Author</u>:

3. <u>Thesis Title</u>:

4. <u>Thesis Committee Members</u>
   <u>Signatures</u>:

5. <u>Distribution Statement</u>: See distribution statements A-X on reverse, then circle appropriate distribution statement letter code below:

   (A) B C D E F X   SEE EXPLANATION OF CODES ON REVERSE

   If your thesis does not fit into any of the above categories or is classified, you must coordinate with the classified section at CARL.

6. <u>Justification</u>: Justification is required for any distribution other than described in Distribution Statement A. All or part of a thesis may justify distribution limitation. See limitation justification statements 1-10 on reverse, then list, below, the statement(s) that applies (apply) to your thesis and corresponding chapters/sections and pages. Follow sample format shown below:

EXAMPLE

<u>Limitation Justification Statement</u> /	<u>Chapter/Section</u> /	<u>Page(s)</u>
Direct Military Support (10) /	Chapter 3 /	12
Critical Technology (3) /	Section 4 /	31
Administrative Operational Use (7) /	Chapter 2 /	13-32

Fill in limitation justification for your thesis below:

<u>Limitation Justification Statement</u> / <u>Chapter/Section</u> / <u>Page(s)</u>

_____ / _____ / _____
_____ / _____ / _____
_____ / _____ / _____
_____ / _____ / _____
_____ / _____ / _____

7. MMAS Thesis Author's Signature: _____

www.ingramcontent.com/pod-product-compliance
Lightning Source LLC
Chambersburg PA
CBHW081257170426
43198CB00017B/2821